Fluid Dynamics of Turbulent Fluidized
Beds for Geldart's Group B Particles

Fluid Dynamics of Turbulent Fluidized Beds for Geldart's Group B Particles

**Vom Promotionsausschuss der
Technischen Universität Hamburg**

zur Erlangung des akademischen Grades

Doktor-Ingenieur(in) (Dr.-Ing.)

genehmigte Dissertation

von
Tom Wytrwat

aus
Rostock

2022

Bibliografische Information der Deutschen Nationalbibliothek
Die Deutsche Nationalbibliothek verzeichnet diese Publikation in der
Deutschen Nationalbibliografie; detaillierte bibliographische Daten sind im Internet
über http://dnb.d-nb.de abrufbar.
1. Aufl. - Göttingen: Cuvillier, 2022
   Zugl.: (TU) Hamburg, Univ., Diss., 2022

1. Gutachter: Prof. Dr.-Ing. Dr. h.c. Stefan Heinrich
2. Gutachter: Prof. Dr.-Ing. Michael Schlüter
Vorsitzender des Prüfungsausschusses: Prof. Dr. Alexander Penn

Tag der mündlichen Prüfung: 18. Januar 2022

© CUVILLIER VERLAG, Göttingen 2022
   Nonnenstieg 8, 37075 Göttingen
   Telefon: 0551-54724-0
   Telefax: 0551-54724-21
   www.cuvillier.de

   ISBN 978-3-7369-7571-2
   eISBN 978-3-7369-6571-3

Abstract

Detailed understanding of the local flow structure is essential for predicting the process behavior of a fluidized bed. To enable the scale-up of processes such as chemical looping combustion, this work provides an in-depth experimental study of the fluid dynamics of turbulent fluidized beds with particles of Geldart's group B.

For this purpose, experiments were carried out in fluidized beds of different diameters in the range of $0.05\,\mathrm{m}$ to $1\,\mathrm{m}$. The superficial gas velocity was varied from close to the minimum fluidization velocity up to $5\,\mathrm{m\,s^{-1}}$ to cover a flow range from the bubbling, over the turbulent, into the fast fluidization regime. Various bed materials, mostly belonging to Geldart's group B, have been characterized and investigated in the experiments with Sauter mean diameters ranging from $13\,\mathrm{\mu m}$ to $348\,\mathrm{\mu m}$ and particle densities between $2600\,\mathrm{kg\,m^{-3}}$ and $4470\,\mathrm{kg\,m^{-3}}$.

Pressure fluctuation analysis has been used as an identification tool for the turbulent fluidized bed regime. Different fluidization conditions have been investigated by variation of the fluidized bed properties, the bed material properties, and the gas properties. The transition velocity from bubbling to turbulent fluidization has been found to strongly depend on the different parameters, which agrees with findings in literature. Based on the measured data, a correlation predicting the transition velocity from bubbling to turbulent fluidization was introduced.

In order to investigate the local flow structure, capacitance probe measurements have been carried out. Using this measurement technique, local solids concentrations and properties of rising bubbles have been determined. Radial profiles of the solids concentration pointed out preferred flow paths of bubbles in dependence of the fluidized bed diameter and the distance to the gas distributor. Furthermore, the vertical size of bubbles was found to continuously grow throughout the bubbling into the turbulent fluidization regime. Nevertheless, a change in bubble shape was concluded from the data. By transition into the turbulent fluidized bed regime the bubble shape is assumed to change from a typical spherical cap into a vertically stretched shape.

Finally, an empiric fluid dynamic model was developed using the local measurement data. It is mainly based on capacitance probe measurements and shows high accuracy in comparison to the pressure data.

Contents

List of Symbols

Latin Symbols

a	exponential decay factor	m^{-1}
a_b	specific inter-phase mass transfer area	$\mathrm{m}^2\,\mathrm{m}^{-3}$
A	cross-sectional area	m^2
A_0	cross-sectional area of a fluidized bed	m^2
b_1	fitting parameter in equation 2.17	$-$
b_2	fitting parameter in equation 2.17	$-$
c_D	drag coefficient	$-$
c_V	solids concentration	$\mathrm{m}^3\,\mathrm{m}^{-3}$
$\overline{c_V}$	mean solids concentration	$\mathrm{m}^3\,\mathrm{m}^{-3}$
$c_{V,\infty}$	solids concentration above the transport disengaging height	$\mathrm{m}^3\,\mathrm{m}^{-3}$
C	concentration	$-$
$d_{3,2}$	Sauter mean diameter	m
$d_{X,3}$	particle diameter at $X\,\%$ in the cumulative distribution	m
d_{ch}	distance of capacitance probe channels	m
d_p	particle diameter	m
D	fluidized bed diameter	m
$D_{i,bm}$	backmixing coefficient of component i	$\mathrm{m}^2\,\mathrm{s}^{-1}$
$D_{i,r}$	radial dispersion coefficient of component i	$\mathrm{m}^2\,\mathrm{s}^{-1}$
$D_{i,z}$	axial dispersion coefficient of component i	$\mathrm{m}^2\,\mathrm{s}^{-1}$
D_{mg}	gas phase molecular diffusivity	$\mathrm{m}^2\,\mathrm{s}^{-1}$
$D_{s,z}$	axial solids dispersion coefficient	$\mathrm{m}^2\,\mathrm{s}^{-1}$
E	arithmetic mean value	$-$
$E_{s,z}$	effective axial solids dispersion coefficient	$\mathrm{m}^2\,\mathrm{s}^{-1}$
f_g	gas friction factor	$-$
f_s	solids friction factor	$-$
f_{Pe}	parameter in eq. 2.38	$-$
G_s	solids circulation rate / entrainment rate	$\mathrm{kg}\,\mathrm{m}^{-2}\,\mathrm{s}^{-1}$
h	variable height	m
Δh	variable height difference	m
H	height	m

H_0	static bed height	m
k_{bs}	inter-phase mass transfer coefficient	m s^{-1}
K	parameter in eq. 2.18	–
K_e	dielectric constant	–
K_i^*	elutriation rate of fraction i	kg m^{-2} s^{-1}
l_p	pierced length	m
L	length	m
m	mass	kg
Δm	mass difference	kg
$\dot{m}$	mass flow	kg s^{-1}
n	parameter in eq. 2.29	–
N	parameter in eq. 5.16	m s^{-1}
P	pressure	Pa
ΔP	pressure drop	Pa
q_3	volume probability density distribution	mm^{-1}
Q	volume flow	m^3 s^{-1}
Q_3	volume cumulative distribution	%
r	radial distance	m
r_{exp}	fluidized bed expansion ratio	m m^{-1}
R	radius	m
R_{corr}	correlation coefficient	–
R_{ij}	reaction term	s^{-1}
R_P	autocorrelation function	–
s	scaling factor in eq. 3.9	–
S	arithmetic standard deviation	–
S_P	power spectral density	mbar Hz^{-2}
t	time	s
Δt	time difference	s
T	temperature	°C
U	velocity	m s^{-1}
U_c	transition velocity from bubbling to turbulent fluidization	m s^{-1}
U_k	transition velocity from turbulent to fast fluidization	m s^{-1}
U_V	voltage	V
$U_{V,0}$	voltage at base capacitance of vacuum	V
V	volume	m^3
x	parameter in eq. 3.9	–
x_i	mass fraction of i	kg kg^{-1}
z	height above gas distributor	m

Greek Symbols

β	fitting parameter in eq. 3.11	—
γ	exponent in section 4.2.3	—
ϵ	bed porosity	$\text{m}^3\,\text{m}^{-3}$
η	dynamic viscosity	Pa s
θ_w	bubble wake angle	$^\circ$
κ	fitting coefficient in equation 6.18	—
μ	mean value	—
ρ	density	kg m^{-3}
σ_x	standard deviation of parameter x	—
τ	time step	s
ϕ	phase hold-up	$\text{m}^3\,\text{m}^{-3}$
$\psi_{c_{V1}c_{V2}}$	cross-covariance function	—
ψ_S	hold-up of suspension sub-phase	—
Ψ	shape factor	—
Ψ_{Wa}	shape factor acc. to Wadell	—
ω	frequency	Hz

Subscripts

0	superficial, initial
abs	absolute
acc	acceleration
b	bulk
bed	standpipe bed
B	bubble
calc	calculated
cap	capacitance probe
D	dense zone
exp	expanded
f	fluid
fa	frequency analysis
fb	fixed bed
fr	frontal
FB	fluidized bed
fric	friction
gb	gas in bubble phase

gs   gas in suspension phase
hyd   hydro dynamic
i   fraction/component
int   interstitial
LS   loop-seal
max   maximum
mf   minimum fluidization
p   particle
ref   reference
s   solid
S   suspension
sc   short-circuiting
se   critical solids entrainment
SP   standpipe
SR   solids return
ss   solids in suspension phase
std   standard deviation
t   terminal
tr   transport
V   volume equivalent
vl   vertical length

Abbreviations

CCS   Carbon capture and storage
CFB100   Circulating fluidized bed with a diameter of $0.1\,\mathrm{m}$
CFB400   Circulating fluidized bed with a diameter of $0.4\,\mathrm{m}$
CLC   Chemical looping combustion
CP1-CP3   Capacitance probe 1-3
ECT   Electric capacitance tomography
FB100   Fluidized bed with a diameter of $0.1\,\mathrm{m}$
FB1000   Fluidized bed with a diameter of $1\,\mathrm{m}$
FB50   Fluidized bed with a diameter of $0.05\,\mathrm{m}$
FCC   Fluid catalytic cracking
MTG   Methanol to gasoline
MTO   Methanol to olefins
MRI   Magnet resonance imaging
OC   Oxygen carrier

PDF	Probability density function
PTC	Positive temperature coefficient
SI	International system
TDH	Transport disengaging height

Constants

g gravitational acceleration $9.81\,\mathrm{m\,s^{-2}}$

Dimensionless Numbers

Ar	Archimedes number (eq. 2.4)
Fr_c	Froude number at transition from bubbling to turbulent fluidization (eqs. 2.18, 2.20)
$Pe_{g,bm}$	Péclet number for gas backmixing (eq. 2.36)
$Pe_{g,r}$	Péclet number for radial gas dispersion (eq. 2.40)
$Pe_{g,z}$	Péclet number for axial gas dispersion (eqs. 2.35, 2.37, 2.38, 2.39)
$Pe_{s,z}$	Péclet number for axial solids dispersion (eq. 2.41)
Re	Reynolds number (eqs. 2.3, 3.7)
Re_c	Reynolds number at transition from bubbling to turbulent fluidization (eqs. 2.16, 2.17, 4.1)
Sc	Schmidt number

1 Introduction

Climate change has become a major issue for nowadays generations. The continuously increasing emissions of greenhouse gases cause growing problems. Amongst others, these are sea-level rise, global temperature rise, warming oceans, shrinking ice sheets, declining arctic sea ice, glacial retreat and ocean acidification due to the absorption of carbon dioxide [1]. Climate change is driven by the emission of various greenhouse gases such as carbon dioxide (CO_2), methane (CH_4), nitrous oxide (N_2O) and much more [2]. With an annual emission of 36.2 Gt in 2017 [3] and a fraction of 72 % of all greenhouse gases [4], CO_2 is the most important driving force. Around 90 % of the CO_2 released by human activities comes from combustion of fossil-fuels and production of cement [3].

To reduce the emissions of greenhouse gases different new technologies have been developed and implemented for energy generation. Some of them are carbon capture and storage (CCS) technologies. In combustion processes using CCS technologies fuels are still burned but the produced CO_2 is captured separately and not released into the atmosphere. Subsequently, the CO_2 can be stored in CO_2 deposits or used for other processes. CCS technologies are bridging technologies that offer the further usage of fossil-fuels without or with reduced greenhouse gas emissions. By this they contribute to the mitigation of climate change.

One of the CCS technologies is the chemical looping combustion (CLC) process. In CLC the CO_2 produced during combustion of carbon containing matter is separated from the combustion air. Thus, it possesses the advantage of the off-gas treatment for extraction of CO_2 being obsolete to reach low carbon emissions. To realize the CLC process two reactors must be coupled. An oxygen carrier (OC) is circulating between these reactors. One of the reactors is passed through by air. The oxygen carrier, which is usually a metal/metal-oxide, is oxidized by the oxygen in this air reactor. Thus, the off-gas of this reactor contains air components only with a reduced fraction of oxygen and can be released to the atmosphere. The oxygen carrier is then transported into the second reactor. Here, the oxygen carrier is reduced and the released oxygen reacts with the fuel. The main components produced during combustion in this fuel reactor are CO_2 and steam. The off-gas of the fuel reactor is treated by oxygen polishing to oxidize other combustion products like carbon monoxide, hydrogen or hydro carbons. Then the water can be condensed in a further process step and CO_2 can be compressed and purified [5]. Finally, the depleted oxygen carrier is recirculated into the air reactor and the process repeats continuously.

Realization of the CLC process usually takes place in a system of interconnected fluidized bed reactors. Plants in different setups have been erected as summarized by Adánez *et al.* [5]. The fluidized bed reactors can be operated in the bubbling, the turbulent and the fast fluidized bed regimes.

Turbulent fluidized beds feature some advantages in comparison to other fluidized bed regimes, such as high solids hold-ups, high heat and mass transfer rates and limited axial gas mixing [6]. With its characteristic properties, the turbulent fluidized bed regime finds – besides CLC – a wide range of application in other processes. The most prominent example is the regenerator of the Fluid Catalytic Cracking (FCC) process [6, 7].

For the construction of large CLC plants it is necessary to find a fluidized bed design that provides high efficiency not only on a laboratory scale but also at industrial scale. To guarantee this, detailed understanding about the fluid dynamic behavior in the fluidized bed and the chemical reaction behavior is necessary. Models are often used for the prediction of the chemical conversion rates. For sufficient predictions, these models must be capable of overcoming the challenges that come with the process scale-up. Fluidized bed reactor models often consist of a fluid dynamic sub-model and one that predicts the complex chemical reaction behavior. Several models describing the fluid dynamic properties of turbulent fluidized beds are available in literature using different approaches [6]. Nevertheless, most of these models have been developed for processes where fine particles are used. These particles mostly belong to group A according to Geldart's classification. In contrast to this, the metal oxides usually used in CLC have larger particle sizes and densities. They show different fluidization behavior than group A particles and are classified as group B. Thus, most of the models for turbulent fluidization in literature give insufficient predictions of fluid dynamic properties like the local solids hold-up. There is a lack of models and information about the fluid dynamic behavior for particles belonging to group B according to Geldart's classification in the turbulent fluidization regime.

For this reason, it is the scope of this work to obtain detailed information about the fluid dynamic behavior of turbulent fluidized beds with particles of Geldart's group B in a thorough experimental study. Furthermore, a fluid dynamic sub-model is developed from the data measured to be able to model a turbulent fluidized bed included into the CLC process.

To predict fluidization behavior sufficiently in laboratory, as well as in large scale facilities, experiments are carried out in fluidized beds of different sizes ranging from 0.05 m up to 1 m in diameter. Superficial gas velocities in the facilities are adjusted to reach states from bubbling, over turbulent, up to fast fluidization. Thereby, regimes are identified and their transitions are quantified. Different bed materials belonging to group B are used with main focus on a fraction of quartz sand having a Sauter mean diameter of 188 µm.

Pressure fluctuations recorded in the different fluidized beds are evaluated. Furthermore, capacitance probes are used and with them solids concentrations, bubble properties and phase hold-ups are determined.

The information gathered from the measurements are used to correlate trends of several characteristic fluidization parameters. Finally, the resulting correlations are used to introduce a model describing the fluidization behavior of turbulent fluidized beds using particles of Geldart's group B.

2 State of the Art

In this chapter an overview about the current state of the art in gas-solid turbulent fluidized beds is given. The different flow regimes and the fluidization behavior of different particles are briefly summarized to provide a classification of turbulent fluidized beds using particles of Geldart's group B. To understand turbulent fluidization, knowledge about the bubbling regime is indispensable, because both flow regimes merge into each other gradually and show similarities. For this reason, a detailed insight into the flow structure of bubbling and turbulent fluidized beds is given in this chapter. In the last section existing literature models for turbulent fluidized beds are summarized.

2.1 Fundamentals of Gas-Solid Fluidization

Fluidized bed technology is widely applied in industrial processes for chemical conversion and particle formulation. Each process requires certain flow conditions and needs to be designed according to the solid material fluidized. Thereby, a fluidized bed can be operated in different states/flow regimes with particles having different fluidization behavior. Both, the flow regimes and particle classification according to their fluidization behavior are discussed in the following.

2.1.1 Flow Regimes

If gas streams through a packed bed of particles at low velocities, the bed induces a pressure drop. This pressure drop increases if the superficial gas velocity is increased. By reviewing information about the flow of fluids through a packed bed of particles, Ergun [8] found the pressure drop to be caused by kinetic and viscous energy losses, which can be expressed by:

$$\frac{\Delta P}{\Delta h} = 150 \frac{(1-\epsilon)^2}{\epsilon^3} \frac{\eta_f U_0}{(d_{3,2}\Psi_{Wa})^2} + 1.75 \frac{1-\epsilon}{\epsilon^3} \frac{\rho_f U_0^2}{d_{3,2}\Psi_{Wa}} \qquad (2.1)$$

The pressure drop ΔP was found to depend on the height difference Δh, the bed porosity ϵ, the gas density ρ_f and viscosity η_f, the superficial gas velocity U_0, the Sauter mean diameter of the particles $d_{3,2}$ (diameter of a particle having the mean volume to surface area ratio of the particles in a bulk) and the shape factor of the particles Ψ_{Wa}.

If the superficial gas velocity is increased beyond a certain point, the particle drag force induced by the gas overcomes the gravitational force acting on the particles. Fluidization begins and the flow of gas changes from a stream through the voids of the stationary particle bed into a state of rising gas bubbles leading to gas induced particle movements. The state of minimum fluidization is reached and the superficial gas velocity at this point is called minimum fluidization velocity U_{mf}. Different approaches for the determination of the minimum fluidization velocity are available in literature [9]. Equation 2.2 gives an approach introduced by Wen and Yu [10]:

$$Re_{mf} = 33.7 \left(\sqrt{1 + 3.6 * 10^{-5} Ar} - 1 \right) \qquad (2.2)$$

It sets the Reynolds number at minimum fluidization Re_{mf} in dependence of the Archimedes number Ar, which are defined by equations 2.3 and 2.4, respectively. ρ_f is the gas density, η_f the gas viscosity, d_p the particle size, ρ_s the particle density, and g the gravitational acceleration:

$$Re_{mf} = \frac{\rho_f U_{mf} d_p}{\eta_f}, \qquad (2.3)$$

$$Ar = \frac{\rho_f \left(\rho_s - \rho_f \right) g d_p{}^3}{\eta_f{}^2} \qquad (2.4)$$

Gas pockets rise characteristically in the shape of bubbles comparable to large gas bubbles rising in liquids, where effects of surface tension and viscosity are small [11]. For this reason, this state is called bubbling fluidization. It is schematically shown in figure 2.1 (b), where the occurrence of bubbles leads to an expansion of the bed in comparison to the fixed bed flow in (a). Whereas the pressure drop over the bed increases steadily in the fixed bed flow, it is constant in the bubbling regime. The pressure drop ΔP of a fluidized bed is induced by the weight of the solid material and can be estimated by [12]:

$$\frac{\Delta P}{\Delta h} = \left(\rho_s - \rho_f \right) \left(1 - \epsilon \right) g \qquad (2.5)$$

Thus, the bed pressure drop depends on the height difference Δh, the solids density ρ_s, the gas density ρ_f, the bed voidage ϵ and the gravitational acceleration g.

The bubbles in a bubbling fluidized bed grow in size with increasing superficial gas velocity and distance from the gas distributor due to coalescence. This can

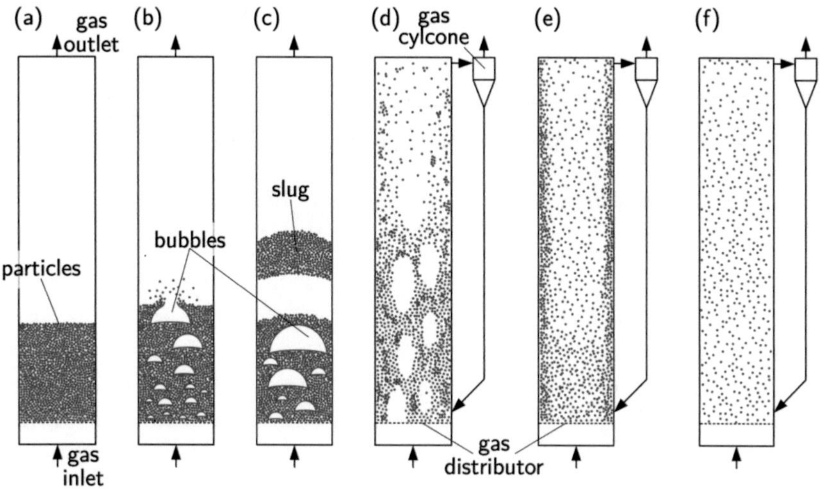

Fig. 2.1: Different fluidized bed flow regimes: (a) fixed bed, (b) bubbling, (c) slugging, (d) turbulent, (e) fast and (f) pneumatic conveying.

be seen in measurements and correlations predicting the bubble size [9, 13]. If bubbles reach the size of the diameter of the fluidized bed, slugs of particles rise periodically in the bed until they break and fall down [9]. The slugging fluidization state is reached, shown in figure 2.1 (c). Slugging does not necessarily appear in a fluidized bed because its occurrence strongly depends on the fluidized bed size and bed height [14].

With further increase of the superficial gas velocity the bubbling or slugging regimes merge into the turbulent fluidized bed regime shown in figure 2.1 (d). A change of the clear bubbly flow structure into a more turbulent, diffuse state occurs and pressure fluctuations, which are mainly induced by bubble rise in the bubbling regime, decrease gradually. In this state of fluidization a bed surface is only barely determinable [9]. Particle entrainment plays a role in turbulent fluidized beds and particles should be recirculated into the bed in continuous processes.

Significant particle entrainment with large rates of solid circulation occurs when the turbulent flow changes into a fast fluidization at larger superficial gas velocities (figure 2.1 (e)) [9]. A characteristic core-annulus flow structure with solids rising in a dilute zone in the bed center and descending at larger concentrations at the fluidized bed wall is formed with particles tending to form aggregates and to move as clusters [12].

If the superficial gas velocity is further increased, pneumatic conveying is reached. In this state particles are transported by the gas in a dilute phase as shown in figure 2.1 (f) [9].

2.1.2 Particle Classification according to Geldart

Because particles of different size and density show different fluidization behavior, Geldart [15] introduced a classification of particles into four different groups:

- Particles belonging to **Geldart's group A** usually have a comparably small mean particle diameter and particle densities below $1400\,\mathrm{kg\,m^{-3}}$. If a fixed bed of these particles is streamed through by gas, it expands considerably before bubbling fluidization sets in. Bubbles in these beds rise faster than the interstitial gas velocity and bubbles coalesce and break-up with a maximum bubble size occurring.

- **Geldart's group B** particles are larger than group A particles and have larger densities. The bed does not expand considerably until reaching the minimum fluidization velocity. Bubbles are known to still rise faster than interstitial gas velocity, whereas there is no evidence for a maximum bubble size for this kind of particles.

- Very fine cohesive powders are difficult to fluidize and are classified to **Geldart's group C**. Gas streaming through these powders tends to form channels due to the inter-particle forces being larger than the forces induced by the gas.

- If particles are very large with high particle densities in comparison to the other groups, they are classified to **Geldart's group D**. In beds of these particles bubbles rise slower than the interstitial gas velocity. Thus, they are streamed through by the gas from the bottom to the top.

According to Geldart [15], these four groups can be differentiated in a diagram of particle size d_p versus particle density ρ_s minus fluid density ρ_f as shown in figure 2.2. The borders between the groups A and B and groups B and D are defined as given by:

$$(\rho_s - \rho_f)\,d_p = 225, \tag{2.6}$$

$$(\rho_s - \rho_f)\,d_p{}^2 = 10^6 \tag{2.7}$$

The border between the groups C and A according to Geldart [15] is based on measurement data.

In addition to Geldart, other authors introduced similar definitions for borders between the four groups [9].

Because this work aims on the investigation of fluid dynamics of turbulent

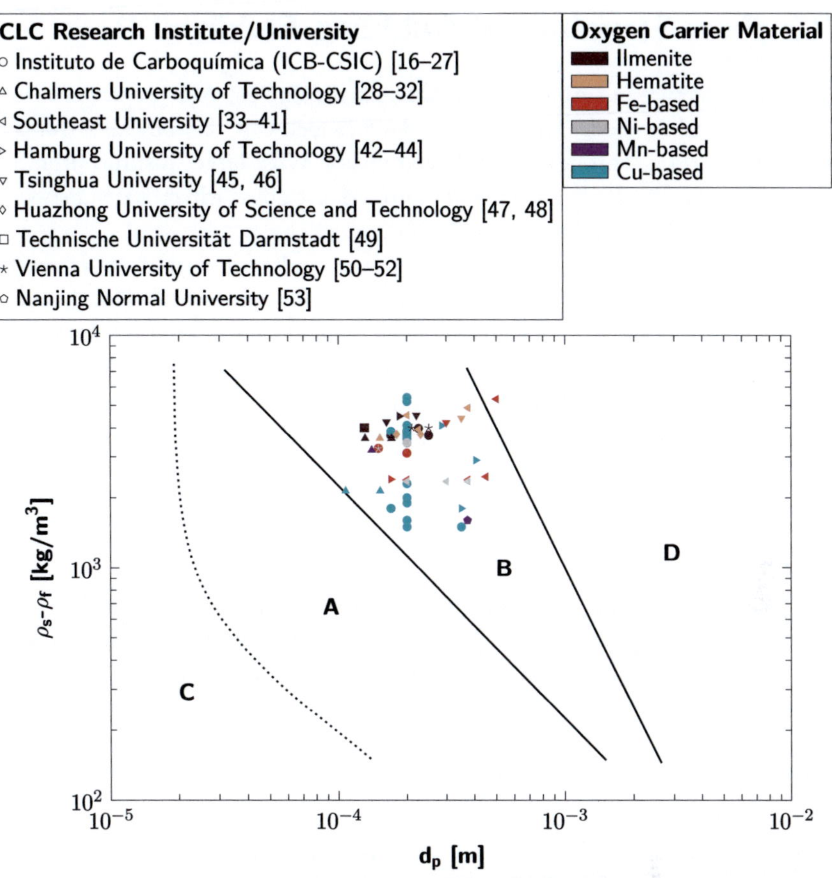

Fig. 2.2: Particle classification according to Geldart [15] with its groups C, A, B and D, its group borders according to equations 2.6 (A-B) and 2.7 (B-D) and different oxygen carriers used in literature for chemical looping combustion (CLC).

fluidized beds as fundamental research for chemical looping combustion (CLC), oxygen carriers used for this process are additionally classified in figure 2.2. These oxygen carriers are based on different metals, which usually have high particle densities. A broad range of densities and particle sizes is used in literature with almost all oxygen carriers clearly belonging to group B according to Geldart's classification (in cases where particle density was not available in the study, information from other studies of the same research group was taken). For this reason, bed materials belonging to Geldart's group B are in the center of interest in this work.

2.2 Flow Structure of Bubbling Fluidized Beds

The flow structure in turbulent fluidized beds is complex. A dense zone exists in the bottom of the bed. There, gas pockets (which can also be called voids or even bubbles) rise through a suspension phase. With larger distance from the gas distributor the dense zone merges into a dilute zone, which is dominated by a flow of particles entrained from the bed.

The transitions from bubbling to turbulent and from turbulent to fast fluidization do not happen by a sudden change in flow structure. The change in flow structure is smooth and points of the transitions can only be determined by definition of certain criteria, which will be explained in detail later. Thus, to fully understand the development of a turbulent flow in a fluidized bed it is indispensable to understand the flow structure and the behavior of bubbles in bubbling fluidized beds.

2.2.1 Generation of Bubbles

Bubbles streaming through the bed are generated at the gas distributor. There are different kinds of gas distributor types commercially available, such as porous plates, perforated plates or bubble cap trays. Depending on the process application, fluidized beds can also be operated with gas injection nozzles. All these different methods of gas distribution/injection into fluidized beds have influence on the generation of bubbles and finally the performance of fluidized bed reactors [11, 54].

A homogeneous gas distribution is preferred for chemical conversion processes to achieve uniform gas-solid contact and also the resulting high heat transfer rates. If gas distribution is inhomogeneous large bubbles can be developed at only one site of the reactor. The main amount of gas inside these bubbles never gets in contact with solid material which decreases conversion rates. To prevent inhomogeneous distribution the pressure drop over the gas distributor should be large compared to bed pressure drop [55].

As described by Karimipour and Pugsley [13], several researchers investigated the initial sizes of bubbles formed at the distributor. Werther and Molerus [56] observed formation of bubbles in small distance above the gas distributor (in this case porous plate) with an initial size of 3-5 mm depending on superficial gas velocity. In a further study, these authors found non-uniform bubble development at the gas distributor [55]. A region of pronounced bubble development close to the wall was observed independent of the gas distributor to bed pressure drop ratio and the size of the fluidized bed. Werther and Molerus [55] explain this behavior with altered packing geometry of particles and different conditions of friction at the wall. In contrast to this, Lim *et al.* [57] ascribe the larger concentration of bubbles at the distributor near the wall to the coalescence process. Bubbles at the wall can only move to the center and probably coalesce or stay at the wall whereas bubbles in the center have more opportunities to move. The result is larger bubbles at the wall close to the distributor.

In fluidized beds with perforated plates or gas injection nozzles the area of active gas injection is mostly small compared to the cross-sectional area of the whole bed. To achieve superficial gas velocities for the bubbling fluidized bed regime or larger the velocity of gas streaming through the orifices of the gas distributor must be much larger. This high gas velocity injection of gas leads to the formation of jets. In contrast to a bubble, where after detachment no void is attached anymore to the distributor, a jet is permanently attached to its gas supplying orifice. Jets can occur permanently (not changing in size), pulsating (oscillating in shape and time with periodical bubble detachment) or even as a spout reaching to the beds surface [58]. A regime map of different jetting regimes was introduced by Penn *et al.* [58]. Their studies were carried out by magnetic resonance imaging (MRI) and deliver a new model for calculation of the jet length. Further approaches for the jet length determination are reviewed by Karimipour and Pugsley [13].

2.2.2 Bubble Properties

After detachment of the bubble from the gas distributor in a fluidized bed it develops a shape different but similar to bubbles in gas-liquid systems. According to Davidson *et al.* [11] bubbles in fluidized beds can be compared to large bubbles in liquids where effects of surface tension and viscosity are small. Despite the fact of the obvious existence of a gaseous and a suspension phase in a fluidized bed, bubbles do not have a surface tension [59, 60]. Thus, the shape of bubbles can be easily disrupted and they can occur irregular in outline and non-uniform in size [61, 62].

2.2.2.1 Bubble Shape and Size

Most bubbles observed in fluidized beds are described as spherical caps as shown in figure 2.3. In their vertical cross section they appear to have a spherical top. To be stable, at the top a bubble must have a pressure large enough to overcome the inertia of the particles and to have the ability to displace them, which results in this form. According to Davidson and Harrison [59] the pressure inside a bubble can be assumed to be homogeneous. The pressure drop in the suspension phase behaves linearly like the pressure drop of a fixed bed. At the top of the bubble the pressure inside is higher than in the suspension phase, whereas at the bottom it is lower. This is shown schematically in the pressure profile in figure 2.3. Thus, at the bottom of the bubble the suspension phase replaces the displaced volume of the bubble and enters the volume of the spherical shape. Particles in this volume are dragged with the bubble and the volume is called wake [55]. The size of the wake is given by the angle θ_w.

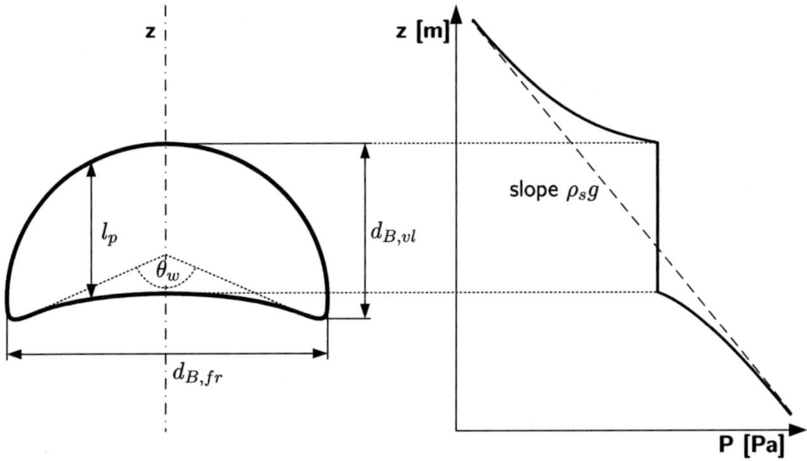

Fig. 2.3: Schematic sketch of the bubble shape with its characteristic dimensions according to Karimipour and Pugsley [13] and the probable pressure profile P along the center axis z of the bubble according to Davidson and Harrison [59].

If a bubble is observed from the top it can be assumed to be spherical in shape. The diameter in this case is called frontal diameter $d_{B,fr}$, which is shown in figure 2.3. This diameter mostly differs from the vertical length of the bubble $d_{B,vl}$. For simplification, bubbles in models are assumed to be spheres having a volume equivalent diameter $d_{B,V}$ of the real bubble. In literature a factor of 1.26 is given to convert the equivalent spherical diameter into the frontal diameter [63].

Bubble sizes are often measured with invasive methods. These methods only allow punctual measurements of concentration variations. The probe enters the passing bubble at a random point at its upper surface. Assuming no lateral movement of the bubble, the probe leaves the bubble at the vertically opposite point at the bottom. The length measured is smaller than the total vertical length of the bubble and is called pierced length l_p. Because it is unknown at which point the probe enters a passing bubble the mean value of the pierced length always comes with a statistical deviation even if only bubbles of same size are measured [64]. Darton *et al.* [63] introduced a factor of 1.6 to calculate the volume equivalent diameter from the measured mean pierced length which is accepted widely in literature [64, 65].

The size of a bubble depends on the properties of the fluidizing medium and the fluidized solid, the plant properties and the distance from the gas distributor. Several authors tried to find correlations for the volume equivalent diameter or the mean pierced length of a bubble in dependence of these different properties. Almost all correlations are empirical or semi-empirical and based on measurements in fluidized beds of different properties. A summary is given by Karimipour and Pugsley [13]. With the application of these correlations measurement conditions always have to be considered. If measurement data for the correlation are based on measurements in 2D fluidized beds e.g. where bubbles are always in contact to walls in one dimension, the results are often not extrapolatable to 3D beds due to different coalescence phenomena and thus different bubble sizes [55, 64, 66]. According to Geldart [66] the bubble diameter in 3D is 1.5 times the diameter in 2D.

Bubble size has a large impact on bed expansion behavior, heat and mass transfer and chemical reaction behavior [67]. It is well known that bubbles in fluidized beds coalesce during their rise from the gas distributor to the bed surface with other bubbles and thus increase in size [55–57, 59, 61, 64, 65, 68]. For particles of group A according to Geldart's classification a maximum bubble size exists [15, 64, 65, 67, 69, 70]. In fluidized beds with these kind of particles inter-particle forces play a significant role due to the small size of particles in this group. The bubble growth is limited and large bubbles break-up. Bubble coalescence and splitting are in equilibrium state, resulting in the aforementioned bubble size maximum [67]. The limitation in bubble size can be advantageous for conversion rates of chemical reactions, because larger bubbles have a lower volume specific surface area and more gas comes into contact with solids [70]. For particles of Geldart's group B and D such a maximum bubble size was not observed [65]. This leads to the assumption that bubbles for these groups are more stable. Furthermore, van Lare *et al.* [64] and Choi *et al.* [68] found only minor influences of particle size on the size of bubbles in group B and D.

The influence of superficial gas velocity U_0 on the bubble size is significant. With increasing superficial gas velocity the size of bubbles increases [61, 64, 66–69, 71]. Geldart found a linear relationship between bubble diameter and the excess gas velocity $(U_0 - U_{mf})$ [66]. This proportionality is reflected by most correlations in literature [13, 67]. An influence of the static bed height on the size of bubbles was not found to be significant [61].

Several applications of fluidized beds are carried out at elevated temperatures and/or pressures. Both parameters mainly affect a change of gas properties.

Thus, these parameters cannot be neglected. For particles of Geldart's group A Hilligardt and Werther [67] found mainly smaller bubbles at higher operation temperatures. This behavior can be explained by a decrease in bubble lifetime (bubbles are less stable) which leads to a reduction in the bubble growth rate [67, 72]. For larger particles a reduction of the size of bubbles at higher temperatures was investigated too. But in this case, the effect was found to be different by Hilligardt and Werther [67]. According to them, the coalescence process changes which leads to a lower bubble growth rate. An increase of pressure was found to reduce bubble sizes as well [72–74]. Čárský *et al.* [73] describe the bubble size as a sophisticated, non-monotonic function of pressure.

2.2.2.2 Bubble Velocity

As mentioned before, bubbles in fluidized beds show similarities to large bubbles in liquids. A single bubble in a fluidized bed $U_{B,single}$ was found to rise with the same velocity as a large bubble in a liquid and is given by [59]:

$$U_{B,single} = 0.71\sqrt{g\,d_{B,V}} \tag{2.8}$$

Here, $d_{B,V}$ is the volume equivalent diameter of the bubble and g the gravitational acceleration. The relationship between the volume equivalent bubble diameter and the single bubble rise velocity comes from a simple force balance of the bubble in the fluidized medium, considering the gravitational force, the buoyancy force, and the drag force of the bubble. This force balance results in

$$U_{B,single} = \sqrt{\frac{4}{3}\frac{1}{c_D}\frac{(\rho_{FB} - \rho_f)}{\rho_{FB}}g\,d_{B,V}}\,, \tag{2.9}$$

with c_D as drag coefficient, the density of the fluidized bed ρ_{FB} and the density of the fluid ρ_f. The drag coefficient for a bubble rising in a fluidized bed has a value of 2.64 [11]. Together with the assumption of $\rho_{FB} \gg \rho_f$ equation 2.9 results in equation 2.8. Equation 2.8 is used by several sources [11, 13] and was validated by up-to-date magnetic resonance imaging (MRI) measurements of Boyce *et al.* [60].

In conventional fluidized beds bubbles do not rise as single bubbles. They are distributed axially and laterally, coalesce and interact with each other. Bubbles in fluidized beds rise in a swarm. The resulting rise velocity of bubbles, the swarm velocity is larger than the velocity of a single bubble rising [59, 75–78]. This increase in bubble velocities is the result of acceleration of bubbles due to coalescence according to Grace and Harrison [75]. Davidson and Harrison [59] introduced equation 2.10 for the bubble velocity U_B, which is the sum of the velocity of a single bubble and the excess gas velocity $(U_0 - U_{mf})$:

$$U_B = U_{B,single} + (U_0 - U_{mf}) \tag{2.10}$$

This approach is widely accepted in literature, but several other correlations are available [13, 75, 78].

The velocity of a bubble depends on its diameter (eq. 2.9). Because the bubble size increases with larger distance from the gas distributor due to coalescence the velocity of bubbles in the bed is also a function of this distance [61, 64, 65]. Furthermore, an increase of the superficial gas velocity influences bubble growth significantly, as explained earlier. This change in bubble growth together with the swarm behavior (eq. 2.10) is responsible for larger bubble velocities [61, 64, 65]. Thus, the bubble velocity is related to the superficial gas velocity.

2.2.3 Coalescence

As described above, coalescence phenomena have large influence on the size and velocity of bubbles. Thus, they decide on the efficiency of a process carried out in a fluidized bed. In literature coalescence is often described to happen between bubbles following each other [11, 75, 79]. According to the pressure profile in figure 2.3, above a bubble, in the cloud, the pressure is larger and in the wake the pressure is lower than if the bubble would be absent. If two bubbles are vertically aligned and get close to each other, their pressure profiles are overlapping. This results in a larger pressure gradient between these bubbles than if they would rise on their own. The increased pressure gradient is responsible for the acceleration of the lower bubble, which leads to a decrease of the distance between both bubbles. Finally, coalescence occurs.

Coalescence does not only take place between vertically aligned bubbles [9, 80, 81]. If two bubbles are horizontally close to each other, they can coalesce as well. Therefore, lateral movements of bubbles are necessary. These movements are mostly induced by the suction of a following bubble into the wake of another [63, 79, 82].

The process of coalescence is driven by both, lateral and vertical movements. It is a result of pressure differences inside the bubbles and their flow fields surrounding them. The probability of coalescence depends on the occurrence of bubbles.

There are several approaches in literature to describe and simplify coalescence processes to get information about bubble sizes and the bed expansion (e.g. [9, 57, 63, 67, 68]). Darton *et al.* [63] introduce a model based on a simple geometric series. This model yields an expression for the bubble diameter and assumes that bubbles rise in preferred paths and bubble growth occurs in a lateral shift to the center of the bed. Another model is introduced by Hilligardt and Werther [67]. They consider bubble growth on the basis of the two-fluid model of Davidson and Harrison [59]. The advantage of the model of Hilligardt and Werther [67] is the applicability to high temperatures. Furthermore, they found by experimental investigation, that coalescence is influenced by a change of temperature.

Most of the coalescence models do not consider the spatial distribution of bubbles. Furthermore, they neglect that at the same height always a broad distribution of bubble sizes exists.

2.2.4 Spatial Distribution of Bubbles

The spatial distribution of bubbles is mainly influenced by coalescence phenomena and the overall particle flow. Coalescence of bubbles does not only lead to larger bubbles but to a decrease of their total number too. This can be measured in vertical direction [83].

2.2.4.1 Lateral Distribution

Laterally, bubbles are not distributed homogeneously. Werther and Molerus [55] found near the gas distributor more bubbles close to the wall than in the center of the bed. They explain this behavior with inhomogeneous bubble generation at the gas distributor. Similar behavior was observed by Lim *et al.* [57]. In contrast to Werther and Molerus [55], they trace back this behavior to coalescence behavior in wall region at the gas distributor. A bubble generated at the wall has different opportunities to move, compared to one generated in the bed center. Obviously, the former has less degrees of freedom in its movement than the latter. Thus, the probability of a bubble at the wall to move laterally into the center is comparably larger. This results in regions of lower concentrations of bubbles near the wall [57].

With increasing height in the bed bubbles coalesce towards the center of the bed. Bubbles near the wall can be rarely found and bubble gas flow in the center increases [55, 57, 63, 64, 69]. The bubbles follow preferred paths of movement. Larger bubbles at the wall close to the gas distributor are forced to move towards the center during rise [55, 63]. On their way these bubbles coalesce with others and create a pair of narrow bands of high bubble concentration, beginning at the bottom near the wall and extending gradually upwards to the center [57]. This flow structure of the bubbles is shown in figure 2.4. Depending on the dimension of the fluidized bed the bands of high concentration form two (or more) circulation cells, as shown in (a) for low aspect ratios, or they can merge in the center and form a stream of large bubbles, as shown in (b) for large aspect ratios (bed height divided by bed diameter) [55, 57]. The formation of the flow structure depends on coalescence behavior and thus on the superficial gas velocity. If large bubbles rise in the center of the bed, a low specific surface area between the bubbly and the suspension phase occurs, in comparison to the bottom. Therefore, the top of a fluidized bed can have a less desirable fluidization quality than the bottom [57].

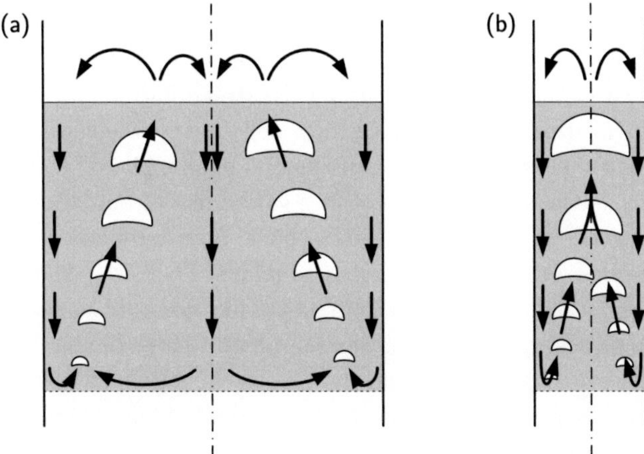

Fig. 2.4: Spatial distribution of bubbles and solid flow paths (arrows) according to Werther and Molerus [55] in (a) beds of low aspect ratios and (b) beds of large aspect ratios.

Bubbles do not only move in preferred paths, they also rise in chains [79]. This phenomenon occurs due to the reduction of drag. A bubble following another bubble experiences a lower resistance in upward moving direction due to the lower amount of solids that have to be displaced compared to a single bubble rising. The reduction of drag leads to an increase in the average bubble rise velocity in a rising swarm of bubbles. Werther developed a model for the rise of bubbles in chains which is able to explain the high coalescence rates in fluidized beds [79].

2.2.4.2 Bubble Frequency

The frequency of bubbles measured in a fluidized bed is directly related to the spatial distribution of the bubbles. For this reason, a radial and axial dependency of the bubble frequency can be measured. Lower frequencies were found with increasing distance from the gas distributor and at the wall [55, 61, 64, 66,

68]. Moreover, the frequency of bubbles increases with increasing superficial gas velocity due to a larger amount of bubbles generated [66, 73, 83]. Information about the influence of the particle size (Geldart group B and D) on the bubble frequency are contradictory in literature. Choi *et al.* [68] and Huang and Lu [74] found the influence to be negligible, whereas van Lare *et al.* [64] and Penn *et al.* [83] describe a decrease of the bubble frequency for larger particles.

2.2.5 Mechanisms of Mixing

Due to the flow of bubbles through the suspension phase particle convection takes place continuously [80]. To describe the streamlines of particles and gas Davidson and Harrison [59] introduced their well known two-phase model based on potential flow theory. The model consists of a bubble phase free of solids and a continuous suspension phase. According to the model bubbles can move as slow bubbles or as fast bubbles in fluidized beds. Slow bubbles rise slower than the gas which passes through the suspension phase. They are passed by the gas as it can be seen in figure 2.5 (a). In contrast to this, fast bubbles rise faster than the gas which passes through the suspension phase. These bubbles displace the suspension phase during their rise and show a gas circulation pattern from inside the bubble into the suspension phase back into the bubble. This is depicted in figure 2.5 (b) [11, 59]. In fluidized beds of fine particles (Geldart's group C, A, B) bubbles rise as fast bubbles due to low minimum fluidization velocities. If larger particles (e.g. Geldart's group D) are fluidized, the occurrence of slow bubbles becomes more probable. The reason for this is the increase of minimum fluidization velocity with larger particle sizes which becomes at one point large enough to overcome the velocity of a rising bubble [67]. Because most processes in fluidized beds are carried out with particles of Geldart's group A and B only the flow structure of fast bubbles will be considered in the following.

2.2.5.1 Particle Movement

Particles are generally dragged with the bubbles and move downwards in regions of lesser bubble activity [55]. The most probable region of particle downflow is the fluidized bed wall. Depending on the dimension of a fluidized bed a downflow of particles can also occur in other regions of the bed, e.g. the center, if the bands of high bubble concentration do not merge as shown in figure 2.4. As long as the rate of particle entrainment is negligible the up-going particle flow must be equal to the flow downwards to fulfill condition of continuity. This overall flow behavior of bubbles in a fluidized bed was found by several authors [55, 57, 84]. Measurements of the particle velocity have proven a mean up flow of

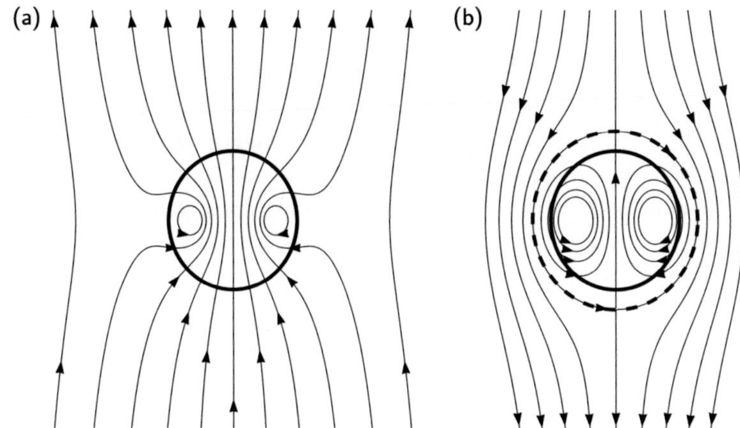

Fig. 2.5: Sketch of fluid streamlines through (a) a slow rising spherical bubble with $U_B < U_0$ and (b) a fast rising spherical bubble with $U_B > U_0$ based on Davidson and Harrison [59] (bubbles are represented as thick solid lines; limit of gas penetration into bubble circulation pattern marked as dashed line in (b)).

particles in regions of high bubble activity with a broad velocity distribution [84]. The distribution of particle velocities broadens with increasing bubble size and according to Penn *et al.* [83] it does not follow a Maxwellian distribution as assumed in granular flow theory. Furthermore, the speed of particles increases with increasing superficial gas velocity and height above the gas distributor due to the occurrence of larger bubbles [83]. In regions where bubble flow does not dominate, particles in the suspension have a Brownian-like motion induced by the gas streaming between the particles [69].

The overall solid circulation induced by bubble rise is not the only effect responsible for mixing. The movement of particles around bubbles, in the wake of a bubble and the eruption at the surface influence the overall mixing behavior [85]. Mixing in fluidized beds takes place rather in vertical than in lateral direction. Macroscopic mixing in lateral direction can be approached by the diffusion equation [86]. The vertical mixing is induced by bubble movement as particle velocity measurements prove. The particle movement and mixing effects can lead to segregation of particles of different sizes or densities. Particles in fluidized beds often have a size distribution and segregation can play a process relevant role. Small particles can segregate into lower layers of the bed and larger particles into upper layers [87]. Detailed information about segregation of binary mixtures in bubbling fluidized beds can be found in [88] and [9].

2.2.5.2 Bubble Wake

A single bubble can carry particles above itself in its cloud or below in its wake [55, 60, 69]. Displaced particles are forming a circulation pattern around the bubble as given by the streamlines in figure 2.5 according to Davidson and Harrison [59, 60]. The fraction of a wake used for modeling in literature is often assumed to have a value of 0.25. This corresponds to a wake-to-bubble volume ratio of 0.33 [80]. According to Huang and Lu [74] the size and thickness of the cloud and wake of a bubble can differ. Larger bubbles have larger clouds and wakes. They carried out their experiments with an invasive two-sensor capacitance probe, where a disruption of the bubble cloud/wake cannot be excluded completely. Rowe and Matsuno [89] found larger wake fractions for larger bubbles because they were found to be proportionally shorter.

Magnetic resonance imaging measurements by Boyce *et al.* [80] showed wake fractions partially similar to the general assumption in literature of 0.25 but most bubbles investigated have not been fully developed due to experimental limitations. Furthermore, measurements of bubbles following each other showed an increase in the size of the wake of the leading bubble. This phenomenon must be considered especially in free bubbling fluidized beds. The wake is not only limited to the extension of the spherical cap to a sphere. If bubbles rise in chains as Werther [79] describes, wakes between these bubbles can extend in size and carry more material than assumed.

2.2.5.3 Gas Backmixing

If at one point in the bed the velocity of particles streaming downwards is larger than the minimum fluidization velocity, the gas flowing through the solids also has a net downflow velocity. Backmixing of gas occurs [55]. The downwards movement of particles and thus also gas mostly occurs in wall-near regions. As mentioned above, in shallow beds with low aspect ratios (static bed height/bed diameter) circulation pattern can develop differently and particle downflow and gas backmixing can also occur in the center [55]. Knowledge about the gas flow pattern and gas concentration distributions is important to quantify mass exchange rates precisely and thus to predict chemical reaction behavior [90]. Further information about gas backmixing in bubbling fluidized beds can be found in [91].

2.2.6 Bed Expansion

Fluidized beds expand due to the gas streaming through them. When particles of Geldart's group A are passed by a gas the bulk already expands before fluidization begins [70]. For group B and D this behavior was not observed. Expansion increases with the formation of bubbles in the fluidized bed. It is important to quantify the expansion of the fluidized bed to get information about heat and mass transfer surface areas [92, 93]. Two parameters for quantification are the bubble holdup ϕ_B and the expanded height H_{exp} of the bed. The bubble holdup of the fluidized bed is defined by equation 2.11. It is zero at minimum fluidization velocity when no bubbles rise in the bed.

$$\phi_B = \frac{V_B}{V_{mf} + V_B} \tag{2.11}$$

Hereby, V_{mf} is the volume of the bed at minimum fluidization and V_B is the volume of all bubbles.

For comparison of expansion behavior of beds with different static bed heights the expansion ratio r_{exp}, the expanded bed height H_{exp} divided by bed height at minimum fluidization H_{mf} in equation 2.12 is used:

$$r_{exp} = \frac{H_{exp}}{H_{mf}} \tag{2.12}$$

Expansion is mainly driven by the bubble volume in the whole bed. The bubble hold-up is a local parameter that changes laterally, axially and temporally as explained before. It is significantly influenced by the superficial gas velocity. A larger superficial gas velocity yields in an increased gas volume that needs to pass the fluidized bed. The bubble volume increases as a result and thus bed expansion increases [65, 75, 83, 92, 94]. Due to larger bubbles erupting at the bed surface at larger superficial gas velocities the standard deviation of the bed expansion also increases [83]. Furthermore, Johnsson *et al.* [92] found the expansion to increase monotonically at low superficial gas velocities. In contrast to this, at larger superficial gas velocities the increase of expansion is reduced.

Because bubbles grow in size with larger distance from the gas distributor bubbles in the upper part of the bed have a larger mean velocity than in the lower part. A larger velocity of bubbles always means a lower residence time. This leads to the assumption that the expansion ratio of a volume element in the upper part of the bed is lower than at the bottom of the bed. For this reason, the static bed height is expected to have an influence on the expansion of the bed. Johnsson *et al.* [92] did not found any influence by bed height, whereas Penn *et al.* [83] found a larger bed expansion for lower static bed height but explained this with a probable measurement error. Generally, the residence time of a bubble in the bed is short. Thus, the change in bed expansion of laboratory fluidized beds is comparably small, which challenges its determination.

Information about the influence of particle size on the expansion behavior is contradictory in literature. While some authors did not find a large impact of particle size [83, 94] others measured a huge influence [65, 92]. Wiman and Almstedt [94] describe an increase of the expansion ratio with increasing pressure in the lower pressure regime, whereas the influence decreases at larger pressures. The temperature was found to have very little influence on the bubble volume fraction and no influence on the expansion of the suspension phase [95]. Thus, the bed expansion is almost unaffected by temperature.

According to the two phase theory the whole gas streaming into the fluidized bed moves upwards in the suspension phase at minimum fluidization velocity and as bubbles. Measurements show that the volume flow of gas rising as bubbles is lower than the difference between volume flow entering at superficial gas velocity and volume flow going through the suspension phase at minimum fluidization velocity. An additional gas flow occurs, which is not provided by the two phase theory. In literature this behavior is explained with short-circuiting of gas between bubbles [57, 65, 66, 75, 83]. Grace and Harrison [75] observed the growth of a bubble at the expense of another one not in contact to it which supports the assumption of gas short-circuiting. According to Shen *et al.* [65] the gas short-circuiting increases linearly with superficial gas velocity and decreases with larger distance from the gas distributor.

The final volume flow balance according to the two phase approach with gas short-circuiting is given by the following equation:

$$Q_0 = Q_{mf} + Q_B + Q_{sc}, \qquad (2.13)$$

where Q_0 is the entering volume flow, Q_{mf} the volume flow going through the suspension phase, Q_B the volume flow through the bubble phase and Q_{sc} the volume flow of short-circuiting gas. To determine the expanded bed height H_{exp} the volume of the bed V_{FB} must be known which is the sum of bubble volume V_B and volume of the suspension phase V_{mf} as given in:

$$V_{FB} = H_{exp} \cdot A_0 = V_{mf} + V_B \qquad (2.14)$$

A_0 is the cross-sectional area of the fluidized bed. The volume of the fluidized bed at minimum fluidization depends on the Geldart's group, as described before. For particles of group B and D it equals to the bulk volume of the bed material. The volume of the bubble phase depends on the volume flow of bubbles and their velocity. As given by equation 2.8 and 2.10 the velocity of bubbles U_B

depends on their size, which in turn is height dependent. This yields:

$$V_B = \int_0^{H_{exp}} \frac{Q_B}{U_B} dh \qquad (2.15)$$

The volume flows of the bubbles and of gas short-circuiting are mostly given by correlations in published models. E.g. Geldart [93] introduced a simple model for bed expansion which uses a correlation based on the Archimedes number for the determination of the bubble volume flow with consideration of the gas short-circuiting flow.

In addition to the bubble volume flow, an approach for the bubble size is necessary for determination of the bed expansion. Therefore, Johnsson *et al.* [92] and Geldart [93] use the bubble size correlation introduced by Darton *et al.* [63], which is based on a coalescence model. It is essential to consider that current models of bed expansion strictly rely on empirical correlations, which have large influences on the results.

2.2.7 Bubble Eruption at the Surface

When a bubble reaches the surface of the bed it erupts. Before a bubble breaks through, particles are carried in the cloud of the bubble above the surface of the bed. The maximum height a bubble reaches above the surface before it breaks through was found to be dependent of bubble size. Larger bubbles push up further above the surface into the freeboard [60]. Particles which were dragged in the wake of the bubble still have a momentum due to their movement with the bubble. Their inertia forces them to get pulled into the freeboard by the bubble [60, 96]. The amount of material and the height particles get pulled into the freeboard depends on process properties. At larger superficial gas velocities larger bubbles occur, which carry more material in their wakes. This material is ejected into the freeboard and due to the decrease of the difference between terminal velocity of the particles and superficial gas velocity it has larger residence times in the freeboard at larger superficial gas velocities. If the bed material has a broad particle size distribution, the terminal velocity of the fines can be smaller than the superficial gas velocity. Thus fines can already be entrained, while the bed is still in the bubbling fluidized bed regime. These fine particles can leave the bed completely, they get entrained.

2.3 Flow Structure of Turbulent Fluidized Beds

By increasing the superficial gas velocity the flow structure, which is dominated by the rise of spherical-cap bubbles in the bubbling regime, changes and the turbulent regime is entered. Bubbles are known to become transient and irregular in shape with indistinct or irregular boundaries [6, 97, 98]. For this reason, they are often called voids in literature to distinguish them from the typical spherical-cap bubble. The change of the flow structure leads to high heat and mass transfer at high solid holdups $(25 - 35\,\text{vol.-\%})$ with limited axial gas mixing [6]. This makes the turbulent fluidization regime favorable for a broad range of process applications like Fluid Catalytic Cracking (FCC), Methanol to Olefins (MTO), Fischer-Tropsch synthesis, acrylonitrile production, silicon chloridization, sulfide ore roasting and Chemical Looping Combustion (CLC) [6, 7, 99, 100].

2.3.1 Break-down of the Bubbly Flow Structure

Several authors conducted detailed analyses of the flow structure of the turbulent regime often using fiber optical probes or capacitance probes for measurement as summarized by Ellis [101] and Bi *et al.* [6]. Here, most studies have been done with particles belonging to group A according to Geldart's classification. There is a general agreement in literature for particles of this group, that the typical two-phase behavior of the bubbling regime, assuming a largely solid empty bubble phase and an suspension phase containing most of the solids, breaks down into a diffuse state, where distinction between two phases becomes impossible [6, 102–106]. This phenomenon happens due to the domination of bubble break-up over coalescence leading to smaller bubbles [106]. As a result the bed appearance becomes more homogeneous [6, 104]. Clift *et al.* [107] introduced a model for the description of the bubble break-up process caused by an indentation in the bubble roof and its growth based on a detailed stability analysis. They found the stability of bubbles to decrease in fluidized beds of larger particulate expansion due to a lower effective kinematic viscosity. Thus, the probability of bubble break-up increases at larger superficial gas velocities overcoming coalescence at some point. Both, break-up and coalescence are probabilistic processes happening in the bubbling regime as well as in the turbulent regime [103]. For this reason the transition between both flow regimes and thus the break-down of the bubbly flow structure occurs gradually [105].

Only few studies exist concerning the flow structure of turbulent beds using particles of Geldart's group B. In addition, contradictory statements were found by different authors. In a study using a fiber optical probe Andreux *et al.* [108] describe the two-phase flow structure to be not applicable anymore by entering the turbulent regime similar to particles of group A. In contrast to this, Lancia *et al.* [109] found the bubbling flow structure to change first into the slugging state to finally become turbulent at larger superficial gas velocities. Their findings might mainly be the result of the static bed height being five times larger than the bed diameter, which usually causes slugging. Furthermore, they found a complete break-down of the two-phase flow structure in the fast fluidization regime. Werther and Wein [110] were able to measure bubble sizes even in the turbulent regime using a capacitance probe. According to them these bubbles cannot be considered as slugs. Through evaluation of the probability density functions of the solids concentrations measured they found the two-phase flow structure to remain in the turbulent regime. In a study comparing FCC particles (Geldart group A) with sand particles (Geldart group B), Bi and Fan [106] found a similar behavior as Werther and Wein [110] for the larger particles, whereas the two-phase flow structure broke down for the smaller ones. This agrees with other studies. Because the break down of the two-phase structure was not found for particles belonging to group B, they describe the flow regimes to bypass the turbulent regime for this kind of particles and instead entering a slugging regime. To distinguish between the different behavior Bi and Fan [106] introduced a demarcation using the Archimedes number of $Ar < 125$ where the two-phase flow structure break down occurs.

2.3.2 Transition from Bubbling to Turbulent Fluidization

The flow structure from bubbling to turbulent fluidization changes in a continuous transition. In literature the transition velocity U_c is usually defined at the maximum of pressure fluctuation amplitudes occurring in the fluidized bed [6]. In a bubbling bed bubbles grow in size with increasing superficial gas velocity. Larger bubble sizes lead to larger pressure gradients along the bubble. Furthermore, larger bubbles displace larger amounts of solids and are responsible for larger eruptions at the bed surface. The result is the increase of pressure fluctuation amplitudes as shown in figure 2.6. By entrance into the turbulent regime the break down of the flow structure into a chaotic state, as described above, occurs [6]. This leads to a maximum in the pressure fluctuation amplitudes and a decrease of this parameter at larger superficial gas velocities. Such a maximum in pressure fluctuations also occurs for particles of Geldart's group B, where a break down of the flow structure might not be the case. Nevertheless, a change of the flow structure takes place as reported in literature [106, 110] leading to a decrease of pressure fluctuation amplitudes when entering the turbulent regime. At larger superficial gas velocities the pressure fluctuations level off. At this point the transition velocity U_k is defined (see figure 2.6) which marks the transition into the fast fluidized bed regime.

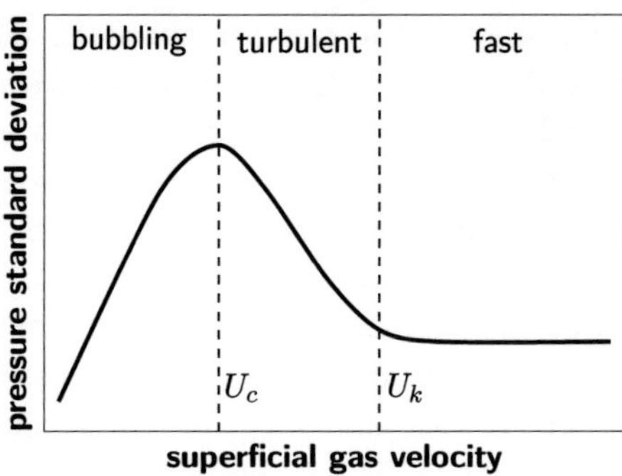

Fig. 2.6: Pressure fluctuations in fluidized beds in dependence of superficial gas velocities and definition of regime transition velocities U_c and U_k.

Several authors, summarized by Bi *et al.* [6], measured the transition velocity U_c in dependence of bed and gas properties. To predict U_c most correlations introduced in literature describe the Reynold's number at the point of transition Re_c (eq. 2.16) as a function of the Archimedes number Ar (eq. 2.4) as shown in the general form in equation 2.17. The latter considers mean particle size d_p and density ρ_s and thus bed material properties. Gas properties are considered by the fluid density ρ_f and dynamic viscosity η_f:

$$Re_c = \frac{\rho_f U_c d_p}{\eta_f}, \tag{2.16}$$

$$Re_c = b_1 \cdot Ar^{b_2} \tag{2.17}$$

The parameters b_1 and b_2 are then fitted to the data measured by the authors.

Using this approach, Seo *et al.* [111] found that their transition velocities measured even at elevated temperatures up to 800 °C can be predicted with sufficient accuracy by their correlation given in table 2.1 (there in eq. 2.23). According to them, the influence of gas properties considered in the Reynold's and Archimedes number satisfy the prediction of U_c. In their work, Seo *et al.* [111] carry out experiments with different size fractions of particles having the same density (Geldart group B) at constant total bed mass (and thus static bed height) and bed diameter. Influences of static bed height [6, 101, 104, 112], bed diameter [6, 101, 104, 112–114], particle size distribution [115] and even measurement position and method [6] are found to strongly influence the results of the transition velocity from bubbling to turbulent fluidization. For this reason, it is questionable whether the approach in equation 2.17 and thus the correlation given by Seo *et al.* [111] is applicable for a wide range of applications and not only for the investigated case.

In addition to the work of Seo *et al.* [111], correlations describing temperature influence on the transition velocity U_c were introduced by Cai *et al.* [113] and Chehbouni *et al.* [114]. Hereby, Chehbouni *et al.* [114] chose an approach similar to equation 2.17 with the Reynold's number being replaced by the Froude number in dependence of the bed diameter as shown in table 2.1 (eq. 2.20). Their experiments were done in two fluidized beds with diameters up to 0.2 m with two different bed materials, one belonging to Geldart group A and the other to B. The correlation derived by Cai *et al.* [113] was developed for a broad range of bed materials. It describes the particle size dependent Froude number in dependence of bed properties and gas properties in relation to ambient conditions as equation 2.18 in table 2.1 gives. In addition to a temperature variation from ambient up to 450 °C, pressure was varied between 1 bar and 8 bar in the study of Cai *et al.* [113] resulting in a detailed view on gas property influences. For particles of Geldart's group B an increase of the transition velocities at larger temperatures is described by all correlations considering temperature dependent data [111, 113, 114]. This behavior was also found in different experimental works [72, 116–118]. For particles belonging to group A contradictory statements have been published. As well as for group B, Cai *et al.* [113] measure an increase of U_c for group A particles at larger temperatures. In contrast to this, Chehbouni *et al.* [114] and Peeler *et al.* [116] found a decrease of U_c for this type of particles.

The influence of static bed height is often directly related to the bed diameter in literature. Dunham *et al.* (acc. to Bi *et al.* [6]) extended equation 2.17 by the aspect ratio of static bed height and bed diameter (table 2.1, eq. 2.19). A similar approach was introduced by Ellis [101] in equation 2.21 in table 2.1 and was further extended later to equation 2.22 [104] under consideration of the bed expansion height instead of the static bed height. Wytrwat *et al.* [112] also extended equation 2.17 to describe the influence of the bed aspect ratio on the transition velocity (table 2.1, eq. 2.25). All authors considering the influence of static bed height and bed diameter found larger transition velocities with increasing aspect ratio. In contrast to this, other authors claim that there is no influence of the static bed height on the transition velocity [6, 119]. The reason for the differences could be the usage of different measurement systems and setups. Bi and Grace [119] describe the usage of a differential pressure measurement system (two-point measurement), whereas other studies like Ellis [101] and Wytrwat *et al.* [112] evaluate absolute pressure measurements or measurements relative to the ambient (single-point measurement).

Larger transition velocities at a constant static bed height are usually measured

Tab. 2.1: Correlations considering influence of particle properties, bed dimensions and/or gas properties on the transition velocity U_c

Author, Year	Correlation	Geldart group	Conditions
Cai *et al.* 1989 [113]	$\dfrac{U_c}{\sqrt{g\,d_p}} = \left(\dfrac{\eta_f\,(20°\mathrm{C})}{\eta_f}\right)^{0.2}\left[K\left(\dfrac{\rho_f\,(20°\mathrm{C})}{\rho_f}\right)\left(\dfrac{\rho_s-\rho_f}{\rho_f}\right)\left(\dfrac{D}{d_p}\right)\right]^{0.27}$ (2.18) $\qquad K = \left(\dfrac{0.211}{D^{0.27}} + \dfrac{0.00242}{D^{1.27}}\right)^{\frac{1}{0.27}}$	A, B	$ambient < T < 450°\mathrm{C}$, $1\,\mathrm{bar} \le P \le 8\,\mathrm{bar}$, $D = 0.15\,\mathrm{m}$.
Dunham *et al.* 1993 acc. to [6]	$Re_c = 1.201 Ar^{0.386}\left(\dfrac{H_0}{D}\right)^{0.128ln(d_p\rho_s)+0.264}$ (2.19)	A, B	$T = ambient$, $D = 0.2\,\mathrm{m}$.
Chehbouni *et al.* 1995 [114]	$\dfrac{U_c}{\sqrt{g\,D}} = 0.463 Ar^{0.145}$ (2.20)	A, B	$ambient < T < 400°\mathrm{C}$, $0.082 < D < 0.2\,\mathrm{m}$.
Ellis 2003 [101]	$Re_c = 0.459 Ar^{0.454}\left(\dfrac{H_{Exp}}{D}\right)^{0.183ln(d_p\rho_s)+0.83}$ (2.21)	A	$T = ambient$, $0.11 < D < 1.56\,\mathrm{m}$, $\frac{H_{Exp}}{D} < 3$.
Ellis *et al.* 2004 [104]	$Re_c = \dfrac{\left(\frac{H_{Exp}}{D}\right)^{0.43} Ar^{0.74}}{\sqrt{4.2\left(\frac{H_{Exp}}{D}\right)^{0.86} + 3.1 Ar^{0.33}}}$ (2.22)	A	$T = ambient$, $0.29 < D < 1.56\,\mathrm{m}$.
Seo *et al.* 2014 [111]	$Re_c = 0.973 Ar^{0.36}$ (2.23)	B	$ambient < T < 800°\mathrm{C}$, $D = 0.078\,\mathrm{m}$.
Rim and Lee 2016 [115]	$Re_c = 0.465 Ar^{0.489}\left(1 + \dfrac{\sigma_{d_p}}{d_{84,3} - d_{16,3}}\right)^{-0.206}$ (2.24)	A, B	$T = ambient$, $D = 0.1\,\mathrm{m}$, $9 < Ar < 2\cdot10^5$, $50 < d_p < 1300\,\mu\mathrm{m}$, $0 < \sigma_{d_p} < 910\,\mu\mathrm{m}$.
Wytrwat *et al.* 2018 [112]	$Re_c = 0.0176 Ar^{0.734}ln\,(230D) + 2.17\dfrac{H_0}{D}$ (2.25)	B	$T = ambient$, $0.05 < D < 0.4\,\mathrm{m}$, $1 \le \frac{H_0}{D} \le 3$.

in beds having a smaller diameter [6, 101, 104, 112, 113]. Studies regarding diameter influence were conducted in beds ranging from small laboratory scale plants up to 1.5 m in diameter as reported by Ellis [101]. The correlation introduced by Chehbouni *et al.* [114] predicts the opposite influencing behavior of the bed diameter. But in contrast to the other studies, Chehbouni *et al.* [114] mainly concentrate on the investigation of temperature influence on U_c and experiments were carried out at only two facilities of comparable small sizes.

As mentioned before, most studies consider the influence of the mean particle size on the transition velocity U_c. Nevertheless, studies about the influence of the broadness of the particle size distribution and the influence of fines are barely available [115, 120, 121]. A correlation based on equation 2.17 considering the standard deviation of the particle size distribution was introduced by Rim and Lee [115] as shown in table 2.1 by equation 2.24. Their results show an increase of U_c if the particle size distribution gets wider at constant mean particle size. Furthermore, they found a decrease of U_c if the amount of fines in the bed is increased.

It can be summarized from literature review regarding the transition velocity from bubbling to turbulent fluidization, that several correlations build on similar approaches exist (as table 2.1 and the review of Bi *et al.* [6] list). Non of these correlations consider influences of all parameters. Furthermore, the differences in measurement methods or systems like single-point/two-point measurements or measurement position, and plant construction like solids return system lead to different results for the transition velocity U_c [6, 119, 120]. In the upper part of a fluidized bed lower transition velocities to turbulent fluidization are measured. They result from a downwards migration of the transition to turbulent fluidization in the bed with increasing superficial gas velocity [119, 120]. Due to different plant sizes and constructions investigated in literature measurement positions (distance from the gas distributor) often vary resulting in incomparable transition velocities.

2.3.3 Bubble/Void Properties

The behavior of bubble properties is well investigated in literature for superficial gas velocities close to minimum fluidization [13]. Despite the applicability of turbulent beds for several processes, studies about the local fluid dynamics and analysis of bubble or void structures and velocities are barely available compared to the bubbling fluidization regime. In addition to this, most studies about the analysis of bubbles in turbulent beds are performed for fluidized beds containing particles of Geldart's group A [6, 101, 103, 122, 123].

2.3.3.1 Bubble/Void Size and Shape

Bubbles/voids in turbulent fluidized beds are generally known to be transient entities with indistinct or irregular boundaries [6, 104]. For this reason, they can be distinguished from clear spherical-cap bubbles occurring in bubbling beds close to minimum fluidization and are often termed void in literature. In this work, "voids" are also called "bubbles" because the transition from one to each other is a continuous process where both, spherical-cap shaped bubbles and transient irregular voids, can coexist and be reshaped into each other.

Due to the reduction of pressure fluctuations at the transition into the turbulent regime the assumption of an increasing influence of bubble break-up leading to smaller bubbles is widely accepted [6]. Supporting this assumption, Zhang and Bi [122] found smaller average pierced lengths than in the bubbling regime by probe measurements of the pierced length (vertical length) of bubbles in turbulent fluidized beds of FCC particles (Geldart group A). A dynamic balance between coalescence and break-up is the result of increasing shearing forces in a turbulent bed due to turbulent eddies. Furthermore, Zhang and Bi [122] found a radial and axial distribution of bubbles that is more uniform in turbulent beds. To predict bubble sizes for these kind of particles they modified the bubble coalescence and splitting model by Horio and Nonaka [124].

Literature about bubble sizes in turbulent beds with larger particles of Geldart's group B is contradicting. Lee and Kim [125] describe a size reduction of bubbles for this kind of particles just as for group A particles, whereas Werther and Wein [110] and Andreux *et al.* [108] found larger bubble sizes even beyond the point of transition into the turbulent regime under similar conditions. Increasing bubble sizes at larger superficial gas velocities in the turbulent regime contradict the increasing importance of break-up phenomena as found for group A particles. In beds of group A particles these break-up phenomena lead to a decrease in the pressure fluctuation amplitudes. For this reason, the reduction of the pressure fluctuation amplitudes in beds of Geldart's group B particles must result from different phenomena like a change of bubble shape. In experiments using magnetic resonance imaging Holland *et al.* [97] showed irregular shapes of bubbles in turbulent beds. Lee and Kim [125] and Werther and Wein [110] assumed a constant shape factor for the conversion of the measured pierced length into the volume equivalent bubble diameter, which is questionable for turbulent beds. Thereby, Werther and Wein [110] used this assumption as an approach for the maximum bubble size only, because the usage of the conversion factor results in bubble sizes close to the fluidized bed diameter. In both studies only vertical bubble sizes could be measured due to the limited information delivered by fibre optical and capacitance probes.

2.3.3.2 Bubble/Void Velocity

Correlations for the determination of the velocity of bubbles in fluidized beds are mostly derived for superficial gas velocities close to minimum fluidization velocity [13]. Usually they describe the velocity of a bubble in dependence of its volume equivalent diameter and the superficial gas velocity as equations 2.8 and 2.10 by Davidson and Harrison [59] show. In turbulent beds these correlations tend to overestimate the bubble velocity due to the change of flow structure, bubble size and shape [6, 108].

For particles of Geldart's group A, Zhang and Bi [122] introduced a predictive correlation similar to Davidson and Harrison [59]. A constant bubble shape factor was assumed by them. The bubble velocity was found to increase linearly with the superficial gas velocity and to be more sensitive to the bubble size than the gas velocity.

In contrast to this, Lee and Kim [125] found almost constant bubble velocities in the turbulent regime for particles of group B, whereas the bubble rise velocities increase at larger superficial gas velocities in the bubbling regime. They modified equations 2.8 and 2.10 to fit their data. Because other authors [108, 110] found contradicting statements for the bubble size trends and Lee and Kim [125] assumed a constant bubble shape even at large velocities, the applicability of their correlation should be clarified.

2.3.4 Voidage Distribution and Bubble/Void Fraction

To describe the flow structure of a fluidized bed it is necessary to distinguish between bed voidage, solids concentration and bubble/void fraction. Usually, the bed voidage ϵ is defined as the volume of fluid V_f divided by the sum of the volume of fluid and volume of solids V_s:

$$\epsilon = \frac{V_f}{V_f + V_s} \tag{2.26}$$

The solids concentration c_V describes the volume of solids divided by the sum of the volume of fluid and solids and thus, the sum of bed voidage and solids concentration is one:

$$\epsilon + c_V = 1 \tag{2.27}$$

Hence, bed voidage and solids concentration consider only the distribution of the fluid and the solids phase. In contrast to this, the bubble or void fraction ϕ_B describes the ratio of the bubble/void phase V_B and the sum of bubble/void and suspension phase V_S as given by

$$\phi_B = \frac{V_B}{V_B + V_S} \tag{2.28}$$

The bubble/void fraction is a parameter only applicable under assumption of the two-phase theory. It can be determined by evaluation of the probability density distributions of the bed voidage or solids concentration which are often measured locally by fiber optical or capacitance probes. Because the local bed voidage in fluidized beds continuously fluctuates, its probability density distribution contains concentrations over the whole range from fixed bed voidage to a value of one. For this reason, a definition of the phases is necessary in order to determine the phase fraction. Different approaches for phase definition in probability density distributions are discussed and summarized in literature [98, 103, 126]. Using these approaches, it must be kept in mind, that the choice of a threshold for phase definition can influence the phase fractions significantly, so that this method can be questionable [6]. Furthermore, Ellis *et al.* [126] found the definition of two phases inappropriate for turbulent fluidized beds with particles of Geldart's group A due to the break down of the two-phase flow structure.

Despite the different definitions of bed voidage and bubble/void fraction both parameters should indicate similar trends if voidage of the suspension phase does not vary significantly in the bed. Because particle entrainment starts to become significant in the turbulent regime in contrast to the bubbling regime, the bed can be divided mainly into a dense bottom zone with high concentrations of solids and a dilute freeboard region above as shown in figure 2.7.

2.3.4.1 Dense Bottom Zone

In turbulent fluidized beds of particles belonging to Geldart's group A, a general increase of the bed voidage was found at larger superficial gas velocities [103, 105]. The radial voidage profile is described being symmetrical with lower voidage (more solids) at the fluidized bed wall and a zone of increased voidage at the bed center. In contrast to this, non-radial symmetric profiles occur in the bubbling regime developing symmetry by increasing the superficial gas velocity [101, 103, 105]. Zhu *et al.* [105] found the voidage at the bed wall to remain constant in the bubbling regime and to increase by entrance into the turbulent regime. Furthermore, they describe large fluctuations of the voidage rather in the center of the bed than at the wall. Close to the gas distributor a not fully developed voidage profile with radial asymmetry can be found even in the turbulent regime [101, 104, 127]. If the superficial gas velocity is increased beyond the turbulent

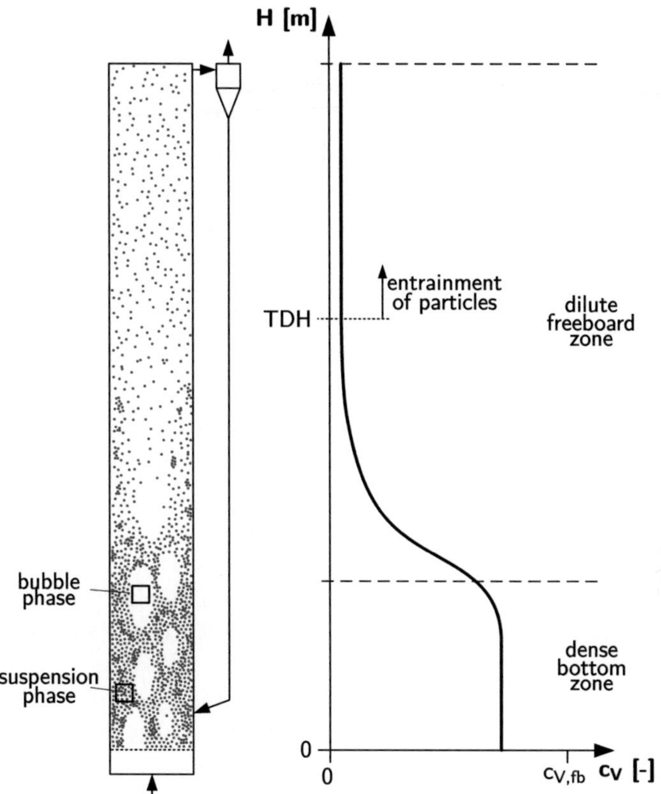

Fig. 2.7: Phases and zones in a turbulent fluidized bed and the resulting mean solids concentration in dependence of the height (TDH = transport disengaging height, see section 2.3.6).

fluidized bed regime, a core-annulus flow structure typical for fast fluidization develops. It is characterized by having a dilute core and an annulus containing large amount of solids [105].

The behavior of the bed voidage in turbulent beds with particles of Geldart's group B is very similar to beds of group A particles. Werther and Wein [110] and Bi and Su [98] also always measured decreasing voidage profiles towards the wall with a zone of increased voidage in the bed center. In contrast to this, in bubbling beds the zone of increased voidage must not lay in the center of the bed necessarily [110]. Furthermore, Andreux *et al.* [108] describe the local radial voidage profiles to keep an almost constant flat shape in the whole turbulent regime. A turbulent fluidized bed can be described as more homogeneous than a bubbling bed [125, 128] with an expansion smaller than it would be predicted by two-phase theory, according to Lee and Kim [128].

Plant design and operating conditions influence the flow structure in a fluidized bed and thus the local bed voidage. Two major parameters that must be considered in scale-up of fluidized beds are the bed diameter and the static bed height. By analysis of the local voidage distribution in beds of different sizes, Ellis [101] found a bimodal distribution at a diameter of 1.56 m and a monomodal distribution at a diameter of 0.29 m at same conditions. This leads to the conclusion that the flow-structure breaks down earlier in smaller beds due to the greater wall effect as it is also found for the transition velocity from bubbling to turbulent fluidization (section 2.3.2). In addition to this, the voidage of the bed is influenced by the static bed height as well [104]. In shallow beds high voidages in the bed center can be found, whereas deeper beds hold larger phase dispersion in this region, with a different circulation pattern [6, 101].

A study comparing two different types of gas distributors was conducted by Sobrino *et al.* [127]. Using a perforated plate they found a more parabolic profile close to the gas distributor with higher voidage in the bed center and a lower one close to the bed wall in comparison to the usage of bubble caps. Furthermore, the position of the solids return is also known to affect the local voidage [101, 119].

The variation of temperature and pressure and thus gas properties is generally known to influence the local flow-structure of fluidized beds [72]. Chehbouni *et al.* [114] measured smaller bubbles and solids aggregates at elevated temperatures by analysis of the local solid fraction fluctuations leading to higher dispersion.

To investigate the influence of the particle size distribution on the bed and single phase voidages, Du *et al.* [103] added a fraction of fines to their bed material (Geldart's group A). No significant influence on the time-averaged cross-sectional bed voidage was found by them.

Additionally to all the process parameters influencing the bed voidage, influences on the results of the voidage measured can arise from the measurement technique used. Ellis [101] describes significant differences in the voidage probability density distributions of capacitance probe and fiber optical probe due to the much larger measurement volume of a capacitance probe. Furthermore, the cross-sectional averaged voidages measured by fiber optical probe agree with the voidages calculated from differential pressure measurements. In contrast to this, Werther and Wein [110] found the capacitance probe to be a reliable tool for the determination of the cross-sectional averaged voidage by validation with γ-ray measurements, whereas pressure gives an overestimation of the solids concentration due to acceleration effects.

Changes in the voidage of a turbulent fluidized bed can also be reached by modification of the standard fluidized bed setup consisting of the turbulent bed having a constant diameter with solids return via a cyclone and loop-seal. An example for this is the circulating turbulent fluidized bed, where secondary fluid is added at a certain distance above the gas distributor to increase solids entrainment beyond this point significantly [100, 129–132]. The result is similar voidage distributions with a reduction in solids backmixing [100, 132] and thus, a flow structure close to plug-flow conditions [129].

For turbulent fluidized beds, different empirical and semi-empirical approaches correlating the time-averaged bed voidage in dependence of the superficial gas velocity, partially considering the radial position, can be found in literature [6, 102, 104, 128]. Thereby, the voidage ϵ of the bed can be represented as given by the Richardson-Zaki-equation [133]:

$$\frac{U_0}{U_t} = \epsilon^n \tag{2.29}$$

According to this equation, the voidage mainly depends on the superficial gas velocity U_0 and the terminal velocity of a single particle U_t. The coefficient n is a system and flow regime dependent parameter [128]. Due to particle interaction in fluidized beds leading to the formation of particle clusters a modified version of equation 2.29 can be used by replacing the terminal velocity of a single particle with the terminal velocity of a cluster [104, 128].

2.3.4.2 Dilute Freeboard Zone

In the freeboard of a turbulent bed a smooth increase of the voidage with increasing distance from the gas distributor can be found [6, 110]. In this region flat voidage profiles compared to the dense bottom zone can be found. These profiles show a slight decrease of the voidage close to the fluidized bed wall, as described by Werther and Wein [110]. According to them, the height dependent decrease of the cross-sectional averaged solids concentration $c_V(h)$ can be described by a simple exponential decay given by

$$\frac{c_V(h) - c_{V,\infty}}{c_{V,D} - c_{V,\infty}} = exp\left(-a\left(h - H_{exp}\right)\right) \tag{2.30}$$

Here, $c_{V,\infty}$ describes the solids concentration above the transport disengaging height (TDH, above this height a constant, height independent solids concentration is reached in the freeboard), $c_{V,D}$ the solids concentration in the dense bottom zone at its expansion height H_{exp} and a the bed material specific decay constant determined by Werther and Wein [110] as given by equation 2.31 in table 2.2.

Tab. 2.2: Correlations for the exponential decay factor a.

Author, year	Equation		Conditions
Werther and Wein 1994 [110]	$aU_0 = \begin{cases} 2.4\,\mathrm{s}^{-1} & \text{for ash,} \quad d_p = 120\,\mu\mathrm{m} \\ 4.0\,\mathrm{s}^{-1} & \text{for sand,} \quad d_p = 240\,\mu\mathrm{m} \end{cases}$	(2.31)	
Adánez $et\ al.$ 1994 [134]	$a\left(U_0 - U_t\right)^2 D^{0.6} = 0.88 - 420 d_p$	(2.32)	Geldart group B, $0.05\,\mathrm{m} \leq D \leq 0.4\,\mathrm{m}$, $60\,\mu\mathrm{m} \leq d_p \leq 1000\,\mu\mathrm{m}$.
Lei and Horio 1998 [135]	$aD = 0.019 \left(\dfrac{G_s}{U_0 \rho_f}\right)^{-0.22} \left(\dfrac{U_0}{\sqrt{gD}}\right)^{-0.32} \left(\dfrac{\rho_s - \rho_f}{\rho_f}\right)^{0.41}$	(2.33)	Geldart group A & B, $0.05\,\mathrm{m} \leq D \leq 0.4\,\mathrm{m}$, $50\,\mu\mathrm{m} \leq d_p \leq 565\,\mu\mathrm{m}$, $700\,\mathrm{kg\,m}^{-3} \leq \rho_s \leq 2600\,\mathrm{kg\,m}^{-3}$, $0.5\,\mathrm{m\,s}^{-1} \leq U_0 \leq 5\,\mathrm{m\,s}^{-1}$.

This approach is widely accepted to describe the decay of the solids concentration, especially in circulating fluidized beds [135–138]. Correlations for the decay constant a are introduced by Adánez *et al.* [134] and Lei and Horio [135] as listed in table 2.2 in equations 2.32 and 2.33. To derive a correlation, Lei and Horio [135] fitted a broad range of literature data of fluidized beds operated with particles belonging to Geldart's group A and B.

2.3.4.3 Suspension Phase Voidage

Knowledge about the voidage in the suspension phase and the related interstitial gas flow through this phase, as well as knowledge about the amount of particles in the bubble phase are important for the prediction of the chemical conversion rate due to their large influence on it. The suspension phase voidage in fluidized beds containing particles of Geldart's group A in the turbulent regime is generally known to expand with increasing superficial gas velocity [6, 98, 102–104]. Chaouki *et al.* [102] describe a linear increase of the suspension phase voidage with increasing superficial gas velocity. By investigation of the influence of particle size distribution on the flow structure, Grace and Sun [120] found a larger expansion of the suspension phase for beds having a wider particle size distribution. According to them, this leads to a larger interstitial gas velocity and a reduced effective viscosity in the suspension phase. Similar results are reported by Du *et al.* [103], who added a fraction of 10 % of fines to the bed material.

For particles belonging to Geldart's group B a similar behavior can be observed. With increasing superficial gas velocity a larger suspension phase voidage and higher interstital gas velocities in the suspension phase can be found leading to a higher rate of bed expansion [108, 110, 125, 128]. Comparing particles of both groups, Bi and Su [98] found the suspension phase voidage of sand particles (group B) to be slightly lower than for FCC particles (group A).

2.3.4.4 Bubble/Void Phase Voidage

Due to the flow behavior and continuous coalescence and break-up processes of bubbles in fluidized beds, particles can always be found in the bubble phase [98]. The voidage of the bubble phase for particles of Geldart's group A is known to increase steadily in the bubbling regime, whereas it levels off by entering the turbulent flow regime [102, 104]. Zhu *et al.* [105] found a clear radial influence on the solids hold-up in bubbles with more solids and thus, lower voidage in bubbles close to the fluidized bed wall in comparison to the center. In addition to this, Grace and Sun [120] detected more solid material inside bubbles when beds with a wide particle size distribution are used in comparison to a narrow one.

In contrast to this, the bubble phase voidage for beds of group B particles is described to increase steadily even in the turbulent regime [108, 139].

2.3.5 Mixing

Mixing in turbulent fluidized beds plays a significant role in plant design and operation [140]. It is responsible for the favorable heat and mass transfer in comparison to other flow regimes and thus it is crucial for chemical conversion in reaction processes [6].

In literature mixing in turbulent fluidized beds is usually described by axial dispersion, backmixing and radial dispersion. Axial dispersion describes the dispersion in a one-dimensional approach without differentiation between longitudinal, or radial dispersion and backmixing [6]. A two-dimensional nonsteady model approach for the description of mixing according to Du *et al.* [7] is given by

$$\frac{\partial C}{\partial t} = D_{i,z}\frac{\partial^2 C}{\partial z^2} + \frac{D_{i,r}}{r}\frac{\partial}{\partial r}\left(r\frac{\partial C}{\partial r}\right) - U_i\frac{\partial C}{\partial z}. \tag{2.34}$$

In this equation the time- (t) and position-depending (axial - z, radial - r) concentration C is described in dependence of the axial and radial dispersion coefficients $D_{i,z}$ and $D_{i,r}$, with i being the phases gas or solids and U_i the phase related velocity. Simplifying this equation by neglecting radial dispersion and assuming steady state, the yielding one-dimensional approach is often used in literature as summarized by Bi *et al.* [6].

Only few studies focus on gas or solids mixing in turbulent fluidized beds, whereas there exist numerous ones for the bubbling flow regime [141–143]. In addition to this, most of the related studies consider the one-dimensional approach only leading to very limited information about radial mixing behavior [6, 7, 141–143].

Gas mixing studies are often carried out by tracer gas experiments. The measurement procedure is not standardized [143]. Nevertheless, the position of tracer injection and its technique have large influence on the measured gas backmixing. This should be considered when using values and correlations from literature for scale-up and plant design [6].

2.3.5.1 Gas Mixing

The behavior of the axial gas dispersion in turbulent beds for particles of Geldart's group A and B is similar. The axial gas dispersion coefficient $D_{g,z}$ follows the curve of the standard deviation of the pressure. In the bubbling regime it continuously increases with the superficial gas velocity, then it reaches a maximum value when the transition velocity from bubbling to turbulent fluidization is reached and it slightly decreases and levels off at superficial gas velocities above the transition velocity [7, 142–144]. The same behavior was found for the gas backmixing coefficient $D_{g,bm}$ which generally behaves like the axial dispersion coefficient due to the radial dispersion being much smaller than backmixing [141, 144].

According to Bi *et al.* [6] axial gas dispersion and gas backmixing are defined in relation to each other together with radial gas mixing. Lee and Kim [141] also distinguish between gas backmixing and axial gas dispersion. They use an one-dimensional diffusion approach for the determination of the gas backmixing coefficient. In contrast to this, Foka *et al.* [143] use the same approach for determining the axial gas dispersion coefficient. Thus, in some cases there is inconsistency in the definition of these coefficients in literature.

The axial dispersion coefficient is usually correlated by its Péclet number $Pe_{g,z}$ in dependence of the system properties as summarized in table 2.3. The Péclet number for axial gas dispersion is defined as given by

$$Pe_{g,z} = \frac{UL}{D_{g,z}}.\qquad(2.35)$$

The Péclet number for gas backmixing $Pe_{g,bm}$ is defined analogical. The velocity U is mostly given by the superficial gas velocity. The characteristic length L can be the bed diameter, the bed radius [141], expanded bed height [145] or the column height [6] depending on the source. The Péclet numbers for axial gas dispersion and gas backmixing are reported to increase monotonously with increasing superficial gas velocity in the turbulent regime [141, 142].

Zhang *et al.* [142] describe the mechanisms behind gas backmixing as a result of inter-phase mass transport of the gas from the bubble phase into the suspension phase and gas being dragged downwards in regions of downwards directed particle flow. For this reason, the internal circulation pattern has a large influence on the gas mixing behavior of the fluidized bed. The local gas backmixing coefficient was measured by Lee and Kim [141]. They found larger backmixing coefficients in the wall region of the bed in comparison to the bed center. In these regions of larger backmixing the residence time distribution of the gas is broader leading to lower conversion rates and a lower reaction selectivity [142].

Most correlations for the axial gas dispersion coefficient are mainly fitted to data of Geldart's group A particles like equations 2.37, 2.38 and 2.39 in table 2.3. A correlation which was explicitly fitted to data of group B was introduced by Lee and Kim [141] with equation 2.36. In their study they also introduced a

Tab. 2.3: Correlations for the axial gas dispersion (g,z) and gas backmixing (g,bm) Péclet numbers.

Author, year	Equation	Conditions
Lee and Kim 1989 [141]	$$Pe_{g,bm} = \frac{U_0 R}{D_{g,bm}} = 2.566*10^{-2} \left(\frac{U_0}{U_{mf}}\right)^{-0.54} \left(\frac{\rho_s}{\rho_f}\right)^{0.067} \left(\frac{d_p}{D}\right)^{-0.588} \quad (2.36)$$	Geldart B, $1.86 \leq U_0/U_{mf} \leq 762$, $892 \leq \rho_s/\rho_f \leq 2076.4$, $1.9 * 10^{-4} \leq d_p/D \leq 3.62 * 10^{-3}$.
Foka $et\ al.$ 1996 [143]	$$Pe_{g,z} = \frac{U_{int} L}{D_{g,z}} = 7 * 10^{-2} Ar^{0.32} \left(\frac{d_p}{D}\right)^{-0.4} \quad (2.37)$$	bubbling & turbulent, $1 \leq Pe \leq 8$, $2 \leq Ar \leq 216$, $510 \leq D/d_p \leq 2667$, $T \leq 600\,°C$.
Bi and Grace 1997 acc. to [145]	$$Pe_{g,z} = \frac{U_0 H_{exp}}{D_{g,z}} = f_{Pe} Ar^{0.32} \left(\frac{D}{d_p}\right)^{0.02344} Sc^{-0.2317} \left(\frac{H_0}{D}\right)^{0.2854} \quad (2.38)$$	$f_{Pe} = 0.247$ for ozone decomposition.
Bi $et\ al.$ 2000 [6]	$$Pe_{g,z} = \frac{U_0 H}{D_{g,z}} = 3.47 Ar^{0.149} Re^{0.0234} Sc^{-0.231} \left(\frac{H}{D}\right)^{0.285} \quad (2.39)$$	fitted mainly to group A data.

Tab. 2.4: Correlation for the radial gas dispersion (g,r) Péclet number.

Author, year	Equation	Conditions
Lee and Kim 1989 [141]	$$Pe_{g,r} = \frac{U_0 R}{D_{g,r}} = 1.472*10^3 \left(\frac{U_0}{U_{mf}}\right)^{-1.012} \left(\frac{\rho_s}{\rho_f}\right)^{-0.446} \left(\frac{d_p}{D}\right)^{0.614} \quad (2.40)$$	Geldart B, $1.33 \leq U_0/U_{mf} \leq 55$, $1245.8 \leq \rho_s/\rho_f \leq 2076.4$, $2.96 * 10^{-4} \leq d_p/D \leq 3.7 * 10^{-3}$.

correlation for the determination of the radial gas dispersion coefficient as shown in table 2.4 by equation 2.40.

Radial gas dispersion was found to be smaller than the axial one by one order of magnitude according to Bi *et al.* [6]. Lee and Kim [141] claim that the difference depends on particle size. In systems of very fine particles radial gas dispersion equals axial, whereas in beds of larger particles the radial gas dispersion is much smaller than the axial one. The behavior of the radial gas dispersion coefficient for particles of group A in dependence of the superficial gas velocity was found to be similar to the axial gas dispersion coefficient with a peak at the transition from bubbling to turbulent fluidization [7]. In contrast to this, for particles of Geldart's group B Lee and Kim [141] describe a continuous increase of the radial dispersion coefficient even in slugging and turbulent fluidization. For this reason, it can be assumed that the radial dispersion coefficient is mainly influenced by bubble size and follows the break-down of the flow structure for group A particles [7], whereas large bubbles can further exist in beds of group B particles in the turbulent regime, as reported in section 2.3.3.1. Vigorous radial mixing takes place in turbulent beds of coarse particles. This was concluded by Lee and Kim [141] from the measured radial dispersion coefficients being much larger than the molecular diffusivity.

As other characteristic parameters of fluidized beds, gas dispersion depends on the fluidized bed system and operation conditions. Most studies are carried out at ambient conditions and quantification of the influences of different parameters is rare [7]. In the range of their investigation, Du *et al.* [7] found only small influences by pressure and temperature on the gas dispersion. Because adding fines to the bed material increases the gross circulation of solids in the bed, radial dispersion is also increased [7]. By comparing baffled turbulent fluidized beds with an unbaffled bed Zhang *et al.* [142] found that the circulation pattern and thus mixing can be influenced significantly. The circulation pattern also changes with bed diameter, as it is known from measurements of the local flow structure. For this reason, the bed diameter influences mixing as well. The axial gas dispersion increases in larger beds as well as with increasing static bed height, as Bi *et al.* [6] summarized.

2.3.5.2 Solids Mixing

Solids mixing in fluidized beds is mainly induced by the movement of bubbles in the suspension which displace particles or drag them in their wake or cloud. Because bubble activity increases with larger superficial gas velocity, the bubble induced solids mixing also increases [7, 140]. This behavior is reflected by the axial and radial solids dispersion coefficients. Most studies concentrate on the determination of the axial solids dispersion coefficient, whereas there is limited information about radial behavior [7]. In a study analyzing mixing of Geldart's group A particles, Du *et al.* [7] found both, the axial and the radial solids dispersion coefficients to increase with rising superficial gas velocity in the bubbling and in the turbulent regime, with a sharper increase in the latter. Similar results were found by Lee and Kim [140], who investigated beds of group B particles. Furthermore, they describe a similarity in the trends of axial solids dispersion and radial gas dispersion which proofs radial gas dispersion to mainly rely on effects of the solids circulation pattern.

The axial solids dispersion is strongly influenced by particle size. Due to larger wake fractions for bubbles in beds of fine particles and the higher mobility of these particles, larger effective axial solids dispersion coefficients are the result in comparison to larger particles [6, 140]. Pressure and temperature were found to have only a small influence on the solids dispersion, compared to gas dispersion [7]. Flow structure and bed behavior changes in larger columns which leads to larger axial solids dispersion coefficients as long as the turbulent eddies increase in size and enlarge the residence time of the solid material [140].

A correlation describing axial solids mixing in a turbulent fluidized bed for particles of Geldart's group B is introduced by Lee and Kim [140] and given in table 2.5 by equation 2.41. To the knowledge of the author of this work, this is the only correlation available for this type of particles at turbulent flow conditions. The Péclet number in this case is defined similar to equation 2.35 with U being the difference of superficial gas velocity and minimum fluidization velocity $U_0 - U_{mf}$, the characteristic length L being the fluidized bed diameter D and the dispersion coefficient being replaced by the effective axial solids dispersion coefficient $E_{s,z} = (1 - \epsilon)\, D_{s,z}$ which depends on the axial solids dispersion coefficient $D_{s,z}$ and the bed voidage ϵ.

2.3.5.3 Inter-Phase Mass Transfer

Inter-phase mass transfer in fluidized beds is dominated by the mixing behavior of solids and gas [6]. A summary of studies about inter-phase mass transfer related to the turbulent fluidized bed regime is given by Bi *et al.* [6]. The mass transfer rate $k_{bs}a_b$ is usually found to increase in the turbulent regime due to the change of flow structure and bubble behavior. Thereby, the volume specific inter-phase transfer area a_b is mainly influencing the mass transfer rate, whereas the mass transfer coefficient k_{bs} tends to be independent of superficial gas velocity [6]. Correlations for the prediction of the mass transfer behavior are given in table 2.6. Bi *et al.* [6] propose the usage of equation 2.42 by Foka *et al.* [143] due

Tab. 2.5: Correlation for the axial solids dispersion (s,z) Péclet number.

Author, year	Equation	Conditions
Lee and Kim 1990 [140]	$Pe_{s,z} = \dfrac{(U_0 - U_{mf})\,D}{(1 - \epsilon)\,D_{s,z}} = 4.222 * 10^{-3} Ar$ (2.41)	bubbling & turbulent, $184 \leq Ar \leq 3.35 * 10^4$, $U_0 \leq 1.3\,\mathrm{m\,s^{-1}}$.

Tab. 2.6: Correlations for the inter-phase mass transfer coefficients.

Author, year	Equation	Conditions
Foka *et al.* 1996 [143]	$k_{bs}a_b = \dfrac{U_0 Sc^{0.37}}{0.613}$ (2.42)	bubbling & turbulent, $U_0 \leq 2.6\,\mathrm{m\,s^{-1}}$, $0.19\,\mathrm{s^{-1}} \leq k_{bs}a_b \leq 4.6\,\mathrm{s^{-1}}$, $0.1\,\mathrm{m} \leq D \leq 0.2\,\mathrm{m}$, $75\,\mu\mathrm{m} \leq d_p \leq 200\,\mu\mathrm{m}$, $570\,\mathrm{kg\,m^{-3}} \leq \rho_s \leq 2650\,\mathrm{kg\,m^{-3}}$, $H_0 = 0.5\,\mathrm{m}$.
Miyauchi *et al.* 1980 [146]	$k_{bs}a_b = 3.7\dfrac{D_{mg}^{0.5}\phi_B}{d_B^{1.25}}$ (2.43)	turbulent, Geldart A.
Zhang and Qian 1997 acc. to [6]	$k_{bs} = 1.74 * 10^{-4} Re_p^{2.14} Sc^{0.81}\dfrac{D_{mg}}{d_p}$ (2.44)	turbulent, Geldart A.

to the large ranges of the parameters investigated in comparison to equations 2.43 and 2.44.

2.3.6 Entrainment

In common processes the bed material consists of particles having different sizes. The terminal velocity of a single particle increases with particle size. By entrance into the turbulent regime the superficial gas velocity can overcome the terminal velocity of fractions of the bed. If these particles reach the freeboard, the zone above the dense bottom zone, they can get entrained (cf. figure 2.7). But because of particle-particle interactions and the formation of particle clusters, reaching the terminal velocity of a bed fraction does not automatically mean they are entrained. Furthermore, the entrainment of larger particles is possible due to the forces induced by smaller entrained particles. The case of particularly entrainment of fines from a bed also containing coarse particles is called "elutriation" [147, 148]. Several correlations for the prediction of the elutriation rate K_i^* are available in literature, summarized by Fotovat [147] and Werther and Hartge [148]. According to Fotovat [147] in turbulent fluidized beds the correlation of Tasirin and Geldart [149] is usually used, which is given by

$$K_i^* = 14.5\rho_f U_0^{2.5} exp\left(-5.4\frac{U_{t,i}}{U_0}\right) \quad \text{for} \quad Re_p > 3000. \qquad (2.45)$$

Thus, the elutriation rate mainly depends on the gas density ρ_f, the superficial gas velocity U_0 and the terminal velocity of a single particle $U_{t,i}$ of the size fraction i. Other factors influencing elutriation are described by Fotovat [147]. By summing up the elutriation rates multiplied by the mass fractions x_i of the particles of same size the entrainment rate G_s can be calculated as given by

$$G_s = \sum_{i=1}^{n} K_i^* x_i. \qquad (2.46)$$

In turbulent beds entrainment becomes significant. For this reason, entrained particles should be recirculated. In this case the entrainment rate equates to the solids circulation rate.

2.3.7 Transition from Turbulent to Fast Fluidization

To distinguish between turbulent and fast fluidization different approaches have been introduced in literature. This is mainly based on the fact that transition occurs gradually without a sudden change in flow structure as it is the case for the transition from bubbling to turbulent fluidization as well. Bi *et al.* [6] summarized three different defined transition velocities:

- transport velocity U_{tr},
- critical velocity U_{se},
- transition velocity U_k

The transport velocity U_{tr} is defined as the velocity at which a sharp change of the pressure gradient does not occur by variation of the solids circulation as it is the case for velocities below. The gradient disappears at velocities above the transport velocity [6].

In contrast to this, the critical velocity U_{se} is based on the onset of significant solids entrainment and can be determined by measurements of the solids circulation rate or the bed emptying time [6].

The transition velocity U_k is defined as the point where pressure fluctuations, as well as the bed voidage level off [6]. This can be seen exemplary in figure 2.6. For its derivation only a pressure measurement system is needed. Therefore, it takes less effort to determine the transition velocity U_k than the transport velocity or the critical velocity.

The transport velocity U_{tr} is usually larger than the critical velocity U_{se}, whereas the latter and the transition velocity U_k are close to each other, indicating that pressure fluctuations level off at the point where significant entrainment begins [6]. Correlations for all three velocities are summarized by Bi *et al.* [6] and are usually correlated by the Reynolds number to the Archimedes number as given by equation 2.17.

2.4 Modeling of Turbulent Fluidized Beds

Modeling of processes conducted in fluidized beds is complex and requires several simplifications. A kinetic and a fluid dynamic model need to be coupled to predict product yield and reaction behavior in the fluidized bed. Because this work aims on the specific fluid dynamics of turbulent fluidized beds, the different kinds of models for this flow regime available in literature are discussed in this part.

Most research on modeling fluidized beds is focused on the bubbling and the fast fluidization regime. Only few authors concentrate specifically on the modeling of turbulent beds [6]. Fluid dynamic models developed for the turbulent regime can be divided into four different main categories:

- single phase plug flow models,

- continuous stirred tank reactor models,
- axially dispersed plug flow models,
- and two-phase models.

Different studies introducing models of these categories are listed in table 2.7. According to Bi *et al.* [6] single phase plug flow models, as introduced by van Swaaij [150] or Bos *et al.* [151], usually over-predict the chemical conversion. In contrast to this, the usage of continuous stirred tank reactor models leads to under-prediction of the conversion [6].

Bos *et al.* [151] introduced a model for the methanol to olefins (MTO) process, which mainly concentrates on the reaction kinetics. A constant solid holdup of 25 vol.-%, ideally mixing of the solids and plug flow of the gas are assumed by them. Reality shows much more complex behavior of turbulent fluidized bed fluid dynamics as described in section 2.3 leading to different conversion of chemicals.

To improve prediction accuracy, some plug flow models include axial dispersion of the gas and thus consider gas backmixing. Edwards and Avidan [152] measured dispersion coefficients in a tracer gas study in beds of different sizes. In their model for the methanol to gasoline (MTG) process they coupled the resulting Péclet numbers with reaction kinetics. The bed was distinguished between dense region and freeboard. Because the model assumes the overall homogeneous appearance of a turbulent bed it requires the suppression of bubbles in reality. For this reason, its applicability is limited to particles of Geldart's group A as it is the case for most models of this type. In turbulent fluidized beds of group B particles the occurrence of large voids is probable [123].

A model similar to the one of Edwards and Avidan [152] was introduced by Sun and Grace [121]. Values for axial dispersion were taken from Guo [165] which were measured for FCC particles. Sun and Grace [121] found a higher accuracy of the model's predictions, if boundary conditions are set to closed at the inlet and open at the top of the bed.

Based on the simple plug flow model of van Swaaij [150], Foka *et al.* [153] introduced a plug flow model with axial dispersion. Dispersion coefficients were measured experimentally for particles of Geldart's group A and B at elevated temperatures leading to a constant Péclet number of 2. Later, the parameter dependency of the Péclet number was further investigated by Foka *et al.* [143] (given by eq. 2.37) and used in the axially dispersed plug flow model of Muradov *et al.* [154].

In two-phase models bubble and suspension phase are considered separately. They were first introduced for bubbling beds under the assumption of the bubble phase being free of solids and the suspension phase being passed by gas at

Tab. 2.7: Turbulent fluidized bed models available in literature.

Type	Geldart group	Author, Year	Comment
single phase	-	van Swaaij 1978 [150]	
plug flow (1D)	-	Bos *et al.* 1995 [151]	
axially dispersed	A	Edwards and Avidan 1986 [152]	
plug flow (1D)	A	Sun and Grace 1992 [121]	
	A, B	Foka *et al.* 1994 [153]	
	A	Muradov *et al.* 2005 [154]	
two-phase (1D)	A	Krambeck *et al.* 1987 [155]	dispersive approach
	A	Sun and Grace 1990 [156]	
	B	Werther and Wein 1994 [110]	only concentrations
	-	Venderbosch 1998 [157]	
	-	Mostoufi *et al.* 2001 [158]	dynamic model
	B	Eslami *et al.* 2012 [159]	
two-phase (2D)	A	Ege *et al.* 1996 [160]	
	A	Farag *et al.* 1997 [161]	
	A	Chaouki *et al.* 1999 [102]	
three-phase (1D)	A, B	Sotudeh-Gharebaagh and Chaouki 2000 [162]	sparger-bubble phase as third phase
	B	Sotudeh-Gharebaagh and Chaouki 2007 [163]	fuel-bubble phase as third phase
combined (1D)	-	Thompson *et al.* 1999 [145]	two-phase (bubbling) + axially dispersion (turbulent)
	-	Abba *et al.* 2003 [164]	two-phase (bubbling) + axially dispersion (turbulent) + core-annulus (fast)

minimum fluidization velocity. These assumptions need to be discussed for turbulent models because their validity is limited to velocities close to minimum fluidization velocity [166].

Two-phase models can be differentiated into one- and two-dimensional models. Two-dimensional models consider axial and radial concentration variations under the assumption of rotational symmetry, whereas one-dimensional neglect the influence of radial variations. Krambeck *et al.* [155] used a one-dimensional homogeneous dispersive approach for modeling the MTG process in a turbulent bed. A countercurrent flow of both phases is assumed, respectively, with mass transfer occurring between the phases quantified by tracer experiments. The model was developed for beds containing particles of Geldart's group A where a dispersive approach of the bubble phase is more suitable than in beds of group B due to the occurrence of relatively small bubbles even at large superficial gas velocities.

Investigations about the influence of particle size distribution on chemical conversion in ozone decomposition was carried out by Sun and Grace [156]. They varied particle size distributions of FCC particles and introduced a one-dimensional two-phase model considering the effects of particle size which appeared to be influencing on several flow regimes with highest prominence between $0.3 - 1.3\,\mathrm{m\,s^{-1}}$ in their experiments.

Further one-dimensional two-phase models were proposed by Venderbosch [157], Mostoufi *et al.* [158] and Eslami *et al.* [159]. The latter model was developed for particles of Geldart's group B. In addition, the model is applicable for elevated temperatures but it was developed and tested for fluidized beds of diameters smaller than $0.1\,\mathrm{m}$, where wall effects are demonstrably strong as explained before.

In addition to the one-dimensional two-phase models named above, Werther and Wein [110] proposed a fluid dynamic submodel predicting axial solids distribution in turbulent beds of group B particles. The model is based on bubble coalescence and growth which in this case is not limited above the transition velocity from bubbling to turbulent fluidization as generally described in literature [6].

Two-dimensional two-phase models for the turbulent regime are barely available in literature and limited to beds of Geldart's group A particles. To consider radial influences, especially the rise of solids in the bed center and the backflow at the wall, Ege *et al.* [160] described the turbulent bed in a core-annulus model as it is usually suitable for the fast fluidization regime [167]. Bubbles are assumed to only rise in the core zone together with solids, whereas the annulus consists of the suspension phase at minimum fluidization condition. In tracer experiments they determined mass transfer coefficients between bubble and suspension phases, as well as between both phases and the suspension in the annulus zone. The mass transfer between bubble phase and suspension phase in core and annulus were found to be significant leading to a high degree of backmixing. In a further study by the same group, Farag *et al.* [161] improved this model.

A different two-dimensional approach was proposed by Chaouki *et al.* [102] to model ethylene synthesis. Using empiric correlations, they describe the bubble phase holdup in dependence of the radius of a fluidized bed facility. Furthermore, the model uses literature correlations for determination of the bubble diameter and velocity. As discussed in section 2.3, correlations for these parameters are mostly determined for superficial gas velocities close to minimum fluidization velocity. For this reason, the applicability of them is questionable in the turbulent regime where, in addition to coalescence, bubble break-up plays a significant role, especially for particles of Geldart's group A.

Additionally, based on the different model types mentioned above, further special models are available in literature. Sotudeh-Gharebaagh and Chaouki [162] introduced a three-phase model for turbulent beds by extending the two-phase theory with an additional bubble phase. Later they modified this model in a further study [163].

An approach of modeling bubbling and turbulent fluidization by coupling a two-phase model and a single-phase axial dispersed plug-flow model was proposed by Thompson *et al.* [145]. In their "Generalized Bubbling Turbulent" model the regime transition is considered to be a continuous process. For this reason, they used a probabilistic approach for determination of the transition. Abba *et al.* [164] further extended this approach by the fast fluidization regime in a "Generic Fluidized Bed Reactor" model.

Only few models in literature proposed for turbulent fluidized beds are applicable to beds containing particles of Geldart's group B, as the overview in table 2.7 shows. The applicability of models developed for group A particles on beds of group B particles is debatable because of the probable large difference in fluid dynamic behavior of both kind of particles in the turbulent regime [6].

3 Materials and Methods

In this chapter the different fluidized bed plants and the bed materials used in this work are presented and characterized. Furthermore, measurement techniques and procedures are explained in detail as well as the evaluation methods used.

3.1 Fluidized Bed Facilities

For the investigation of different plant properties on the fluid dynamic behavior of fluidized beds five plants are used in this work. In the following the facilities and corresponding peripheries are described in detail. All plants have been used in previous published works [112, 123, 168].

3.1.1 Fluidized Bed Plant FB50 (Diameter: 0.05 m)

A fluidized bed with a diameter of 0.05 m called FB50 was used for determination of the influence of bed diameter on regime transition. Its flow sheet is shown in figure 3.1 (a). The plant is made from acrylic glass and operated at ambient conditions. Dry air is supplied via a pressurized air network as fluidization gas. The superficial gas velocity is adjusted using a mass flow controller (F 203AC FA by Bronkhorst High-Tech B.V.) and was set in a range between $0.1\,\mathrm{m\,s^{-1}}$ and $2.2\,\mathrm{m\,s^{-1}}$. The plant is equipped with a porous plate as gas distributor. At a height of 0.8 m above the gas distributor the diameter of the facility is extended to prevent particles from getting entrained. Particles entrained despite the diameter extension are collected by a filter.

Pressure measurements relative to the ambient were carried out in the windbox (pressure transducer: 142PC15D-PCB by First Sensor), at a height of 0.05 m above the gas distributor (pressure transducer: 142PC05D-PCB by First Sensor) and in the extended cross-section of the bed (pressure transducer: 142PC05D-PCB by First Sensor). The pressures have been recorded at a frequency of 100 Hz.

The static bed height was varied between $0.05 - 0.2\,\mathrm{m}$ to reach aspect ratios (static bed height divided by bed diameter) of $1 - 4$.

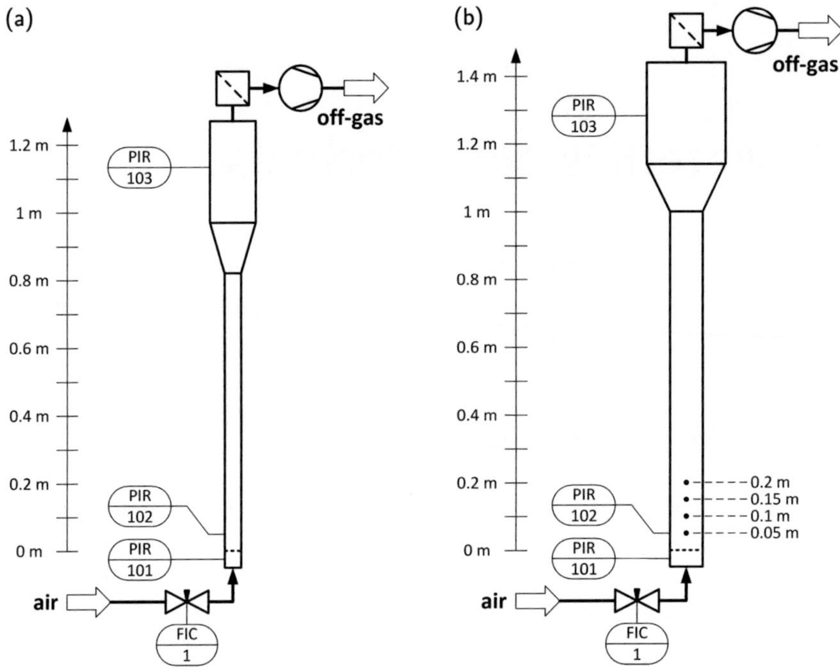

Fig. 3.1: Flow sheets of (a) the FB50 ($D = 0.05\,\mathrm{m}$) and (b) the FB100 ($D = 0.1\,\mathrm{m}$).

3.1.2 Fluidized Bed Plant FB100 (Diameter: 0.1 m)

The fluidized bed plant FB100 has a diameter of $0.1\,\mathrm{m}$ and is made from acrylic glass as well. Its setup is similar to the FB50 as can be seen in figure 3.1 (b). The same gas supply system and mass flow controller are attached. The gas, which is distributed via a porous plate, was set to velocities in a range of $0.1 - 1.7\,\mathrm{m\,s^{-1}}$. At a height of $1\,\mathrm{m}$ the bed cross-section is extended to prevent solids entrainment. Nevertheless, entrained particles are collected by a filter.

Pressure measurements were done at ambient conditions using the same equipment and measurement positions as in the smaller facility FB50. In addition to the pressure measurements, this plant was used for probe measurements. The probe ports are installed at heights of $0.05\,\mathrm{m}$, $0.1\,\mathrm{m}$, $0.15\,\mathrm{m}$ and $0.2\,\mathrm{m}$. To obviate influences from electrostatic charging on probe measurements the inside of the facility was coated and grounded with a layer of alumina foil.

In this facility the static bed height was varied between $0.1 - 0.3\,\mathrm{m}$ to reach aspect ratios between $1 - 3$.

3.1.3 Circulating Fluidized Bed Plant CFB100 (Diameter: 0.1 m)

In addition to the FB100, another fluidized bed plant (CFB100) having a diameter of 0.1 m was used which can be operated as circulating fluidized bed and at temperatures up to 800 °C. The flow sheet of this plant is shown in figure 3.2. Dry air from a pressurized air network is used for fluidization. The air is preheated via an air preheater having a power of 2.5 kW for norm volume flows below $15\,m^3\,h^{-1}$ and one with a power of 7.5 kW for volume flows above. A maximum gas inlet temperature of 500 °C was set to protect the preheater. The air flow was set by a mass flow controller (F 203AC FA by Bronkhorst High-Tech B.V.) in a superficial gas velocity range of $0.2 - 3.5\,m\,s^{-1}$ at experimental conditions. The preheated gas enters the windbox of the fluidized bed and is distributed via a 3 mm thick porous plate made from Hastelloy X (filter grade SIKA HX 20 AX). A drain pipe is welded to the center of the porous plate to enable the complete emptying of the facility. The fluidized bed is made from stainless steel and has a height of 15.6 m. The bed material can be filled into the fluidized bed at a height of 2.5 m above the distributor plate via a feed pipe. Entrained particles leave the fluidized bed via an abrupt exit into a cyclone. The solids are lead back into the fluidized bed through a standpipe and a loop-seal, which is fluidized with a norm volume flow of $7.5\,m^3\,h^{-1}$ of air.

The plant is heated electrically with 14 heating elements for the fluidized bed and the standpipe, respectively. The temperature is monitored and recorded in the windbox, at ten positions in the fluidized bed, and at six positions in the standpipe and loop-seal with Ni-Cr-Ni thermocouples. The installation heights are shown in the flow sheet. Thereby, the exact measurement position is in the center of the fluidized bed.

18 differential pressures are measured with 13 being distributed over the fluidized bed. In addition to this, pressures relative to ambient conditions are measured in the windbox and at a height of 0.28 m above the gas distributor and recorded at a frequency of 100 Hz.

The static bed height was adjusted between $0.1 - 0.3$ m to reach aspect ratios between $1 - 3$.

The gas leaving the cyclone is led into an after burning chamber (not depicted) which is usually used for combustion experiments but was not in operation during experiments of this work. Finally, the gas is cooled, filtered and then sucked out by a suction fan which induces a slight constant under pressure in the fluidized bed.

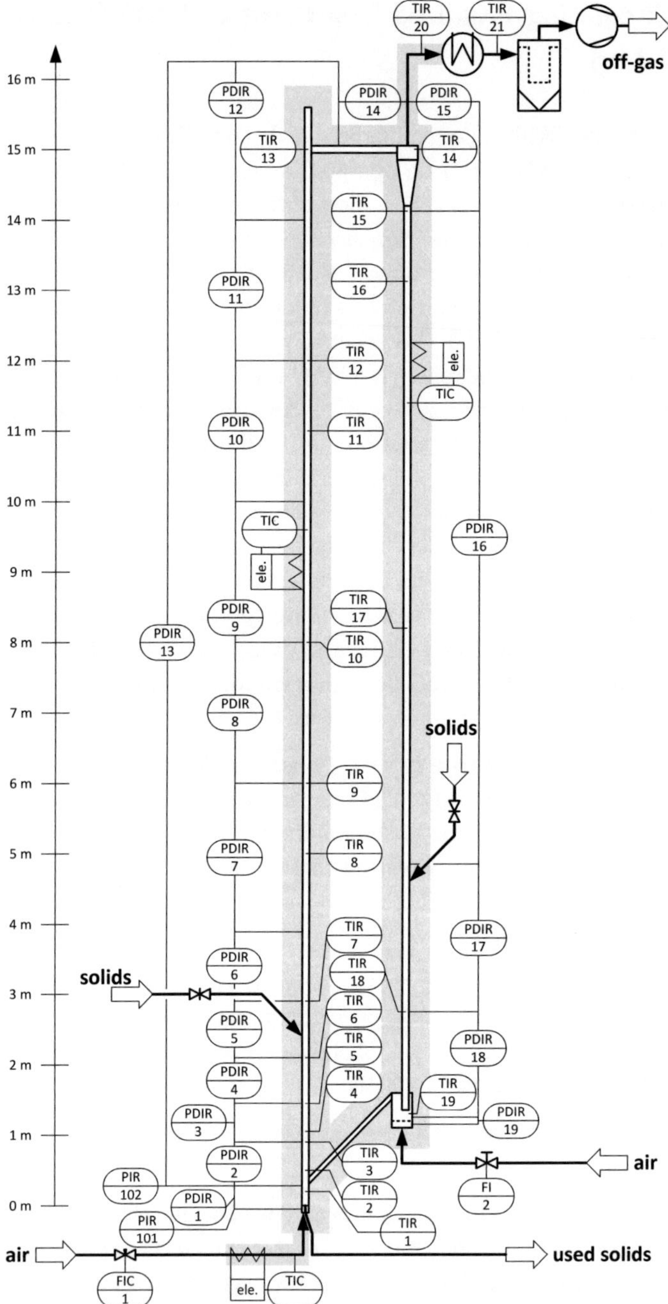

Fig. 3.2: Flow sheet of the high temperature facility CFB100 ($D = 0.1\,\text{m}$).

3.1.4 Circulating Fluidized Bed Plant CFB400 (Diameter: 0.4 m)

The fluidized bed plant CFB400 has a diameter of 0.4 m and is operated at ambient conditions. Its flow sheet is shown in figure 3.3. Two different air blowers were used to realize a large range of superficial gas velocities. The roots blower GMa 11.3 (Aerzen Maschinenfabrik GmbH) was used for the range of $0.2 - 1.0\,\mathrm{m\,s^{-1}}$ and the roots blower GMb 14.9 (Aerzen Maschinenfabrik GmbH) for the range of $1.0 - 5.0\,\mathrm{m\,s^{-1}}$. Measurements of the volume flow are carried out by two orifice flowmeters according to DIN EN ISO 5167-2 [169], one for each air blower respectively. Both air blowers are connected by separate piping systems to the windbox. The air enters the fluidized bed via a porous plate (sintered plate of 300 µm polyethylene particles), which is supported by two perforated metal plates (staggered pitch, hole diameter: 5 mm, hole distance: 8 mm, thickness: 2 mm) below and above. The fluidized bed has a height of 15.6 m from the gas distributor and is constructed from steel segments with four glass elements in between at heights of about 2 m, 6 m, 9 m and 12 m. Gas and entrained particles leave the fluidized bed via an abrupt exit into a cascade of two cyclones. Solids are returned via a standpipe and a loop-seal, which is operated at a norm volume flow of $89.4\,\mathrm{m^3\,h^{-1}}$. The returned solids enter the fluidized bed at a height of 1 m above the gas distributor. For measurement of the solids circulation rate a separately mounted section with a butterfly valve is installed in the standpipe, which is explained in detail later. The solid material is stored in a storage hopper, which is connected to the standpipe. Thus, the fluidized bed can be emptied into or fed from the material storage. The storage hopper is mounted on a load cell to measure the weight of the stored material. The gas leaving the cyclone cascade is filtered and blown to the ambient by a suction fan.

Temperatures inside the plant were measured behind the orifice flowmeter and inside the fluidized bed at a height of 3 m above the gas distributor.

The plant is equipped with a pressure measurement system consisting of 27 differential pressure sensors to monitor the fluidized bed and of ten differential pressure sensors installed in the plant periphery and the standpipe. In addition to this, pressures relative to the ambient are measured at the measuring orifices, in the windbox, 0.25 m above the gas distributor and 10.86 m above the gas distributor. The pressure measuring positions are shown in figure 3.3 with a detailed sketch of the bottom section of the plant in figure 3.4. All pressures are recorded at a frequency of 100 Hz.

Probe measurements are done at eight probe ports distributed over a height of 1.5 m above the gas distributor (0.19 m, 0.26 m, 0.33 m, 0.43 m, 0.53 m, 0.73 m,

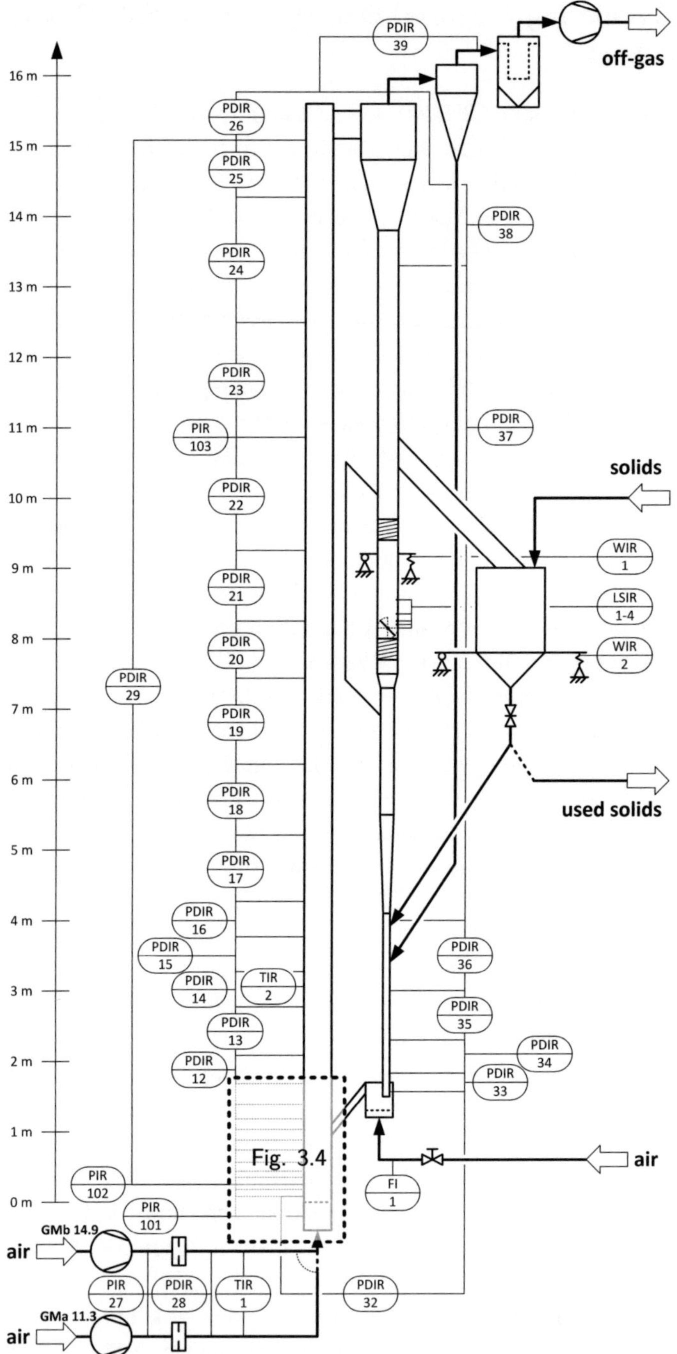

Fig. 3.3: Flow sheet of the CFB400 ($D = 0.4\,\mathrm{m}$).

0.86 m, 1.5 m). To measure the radial and axial symmetric behavior of flow properties, at a height of 0.19 m four probe ports in angles of 45° to each other are installed. The angles of solids return (337.5°), air supply (90°) and the different probe ports are shown in figure 3.4.

The plant was operated in steady state at different superficial gas velocities and different static bed heights. The static bed height was varied between $0.4 - 1.6$ m to reach aspect ratios in a range of $1 - 4$.

3.1.5 Fluidized Bed Plant FB1000 (Diameter: 1 m)

The plant with the largest diameter used for the fluid dynamic investigations in this work has a diameter of 1 m and is called FB1000. The flow sheet is shown in figure 3.5. Air is supplied by two roots blowers for different ranges of superficial gas velocities. The roots blower GMa 13.8 (Aerzen Maschinenfabrik GmbH) was used for a superficial gas velocity range of $0.3 - 0.7\,\mathrm{m\,s^{-1}}$ and the roots blower GMb 16.12 (Aerzen Maschinenfabrik GmbH) for a range of $0.75 - 1.25\,\mathrm{m\,s^{-1}}$. The piping system was reconstructed to change between both blowers. Measurements of the volume flows are carried out with orifice flowmeters (acc. to DIN EN ISO 5167-2 [169]), one for each air blower. To prevent electrostatic charging which influences probe measurements at large fluidized bed diameters, the air was humidified using two steam generators having thermal a power of 9 kW each. The plant is equipped with a porous plate (sintered plate of $200 - 250\,\mathrm{\mu m}$ polyethylene particles) having a thickness of 1 cm which is supported by two perforated metal plates (staggered pitch, hole diameter: 5 mm, hole distance: 8 mm, thickness: 2 mm), one above and one below. The fluidized bed has a height of 4.6 m and is made from steel. Entrained solids and gas leave the plant via an abrupt exit into a cyclone where both arc separated from each other. The solids flow down in a standpipe. Gas leakage through the standpipe is prevented by two butterfly valves that are hand-operated during the experiments. Due to the construction of the cyclone and standpipe being designed for the bubbling fluidized bed regime, superficial gas velocities above $1.25\,\mathrm{m\,s^{-1}}$ could not be realized. Gas leaving the cyclone passes through two filters into a suction fan and is released to the ambient.

Probe ports at four different heights (0.18 m, 0.53 m, 0.88 m, 1.23 m) are used for determination of the local flow structure by capacitance probe. They are installed straight on the opposite site to the air supply. The angles of solid return (30°), probe ports (90°) and air supply piping (270°) are shown in the top view of the facility in figure 3.5.

The plant was operated at two different static bed heights of 0.5 m and 1 m being aspect ratios of 0.5 and 1.

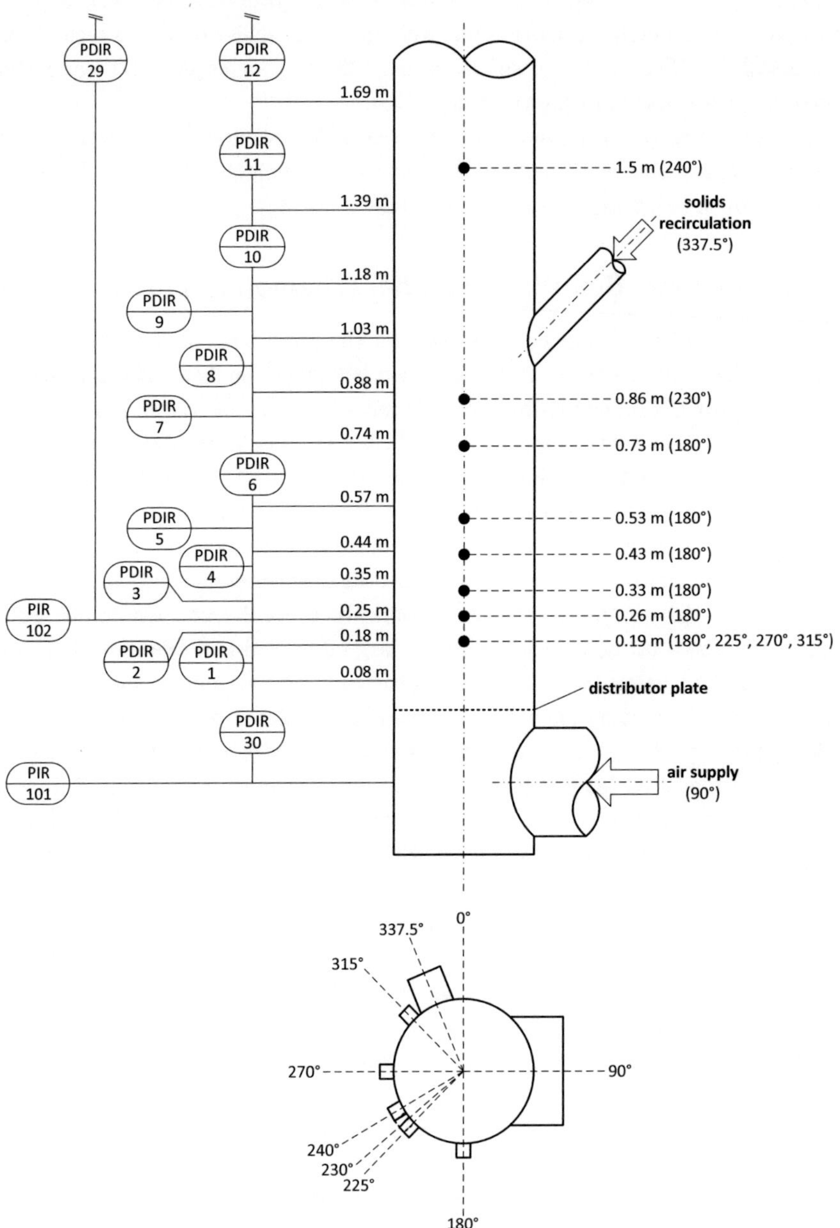

Fig. 3.4: Schematic sketch of the bottom section of the CFB400 in detail including the top view.

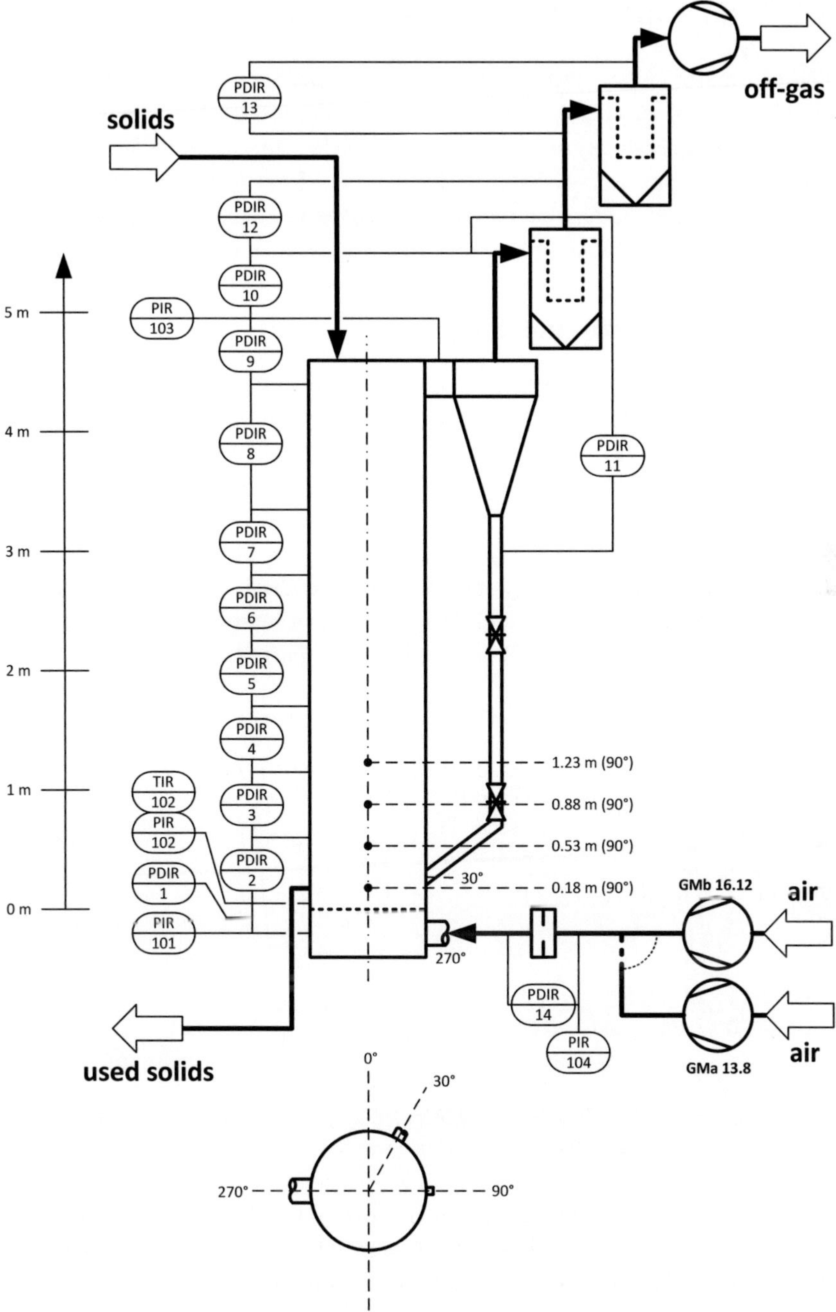

Fig. 3.5: Flow sheet of the FB1000 ($D = 1\,\mathrm{m}$) including the top view of the fluidized bed.

3.2 Bed Material Characterization

The bed materials are chosen according to properties of oxygen carriers used
in literature for the chemical looping combustion process. Major fluid dynamic
parameters of these oxygen carriers are particle size, distribution, shape and
density. In figure 2.2 various oxygen carriers from literature are classified
according to their fluidization behavior belonging to Geldart's group B. For this
reason, the reference material used in this work was chosen accordingly.

3.2.1 Particle Size and Shape

Five different fractions of quartz sand and one fraction of ilmenite were used for
the experiments. The probability density and the cumulative distributions are
shown in figure 3.6 (a) and (b). The Sauter mean diameters $d_{3,2}$ and the median
diameters $d_{50,3}$ are listed in table 3.1 additionally. The particle size distributions
were measured with a Camsizer XT (formerly: Retsch Technology GmbH, since
2020: Microtrac Retsch GmbH). Measurement principle is an dynamic image
analysis method according to the international standard ISO 13322-2 [170]. The
measurement is two dimensional. As parameter of interest the diameter of an
area equivalent circle to the 2D projection of the particle is chosen.

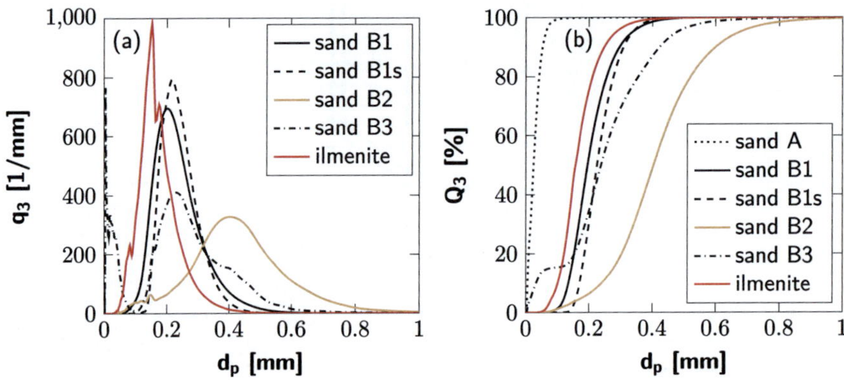

Fig. 3.6: (a) probability density distributions and (b) cumulative distributions of the
particle sizes of the different bed materials used in this work.

Sand B1 was used for experiments in all facilities listed above. Another fraction
having a larger mean particle size than sand B1, called sand B2, was used for
investigations of the transition velocities from bubbling to turbulent fluidization.
For this, the plants having diameters of 0.05 m and 0.1 m have been used at
ambient and high temperature conditions. To determine influences of the particle

Tab. 3.1: Bed material properties: particle sizes and shape factor.

material	$d_{3,2}$	$d_{50,3}$	$\frac{d_{90,3}-d_{10,3}}{d_{50,3}}$	Ψ
	[µm]	[µm]	[−]	[−]
sand A	13	24	1.93	0.796
sand B1	188	196	0.82	0.835
sand B1S	222	226	0.61	0.848
sand B2	348	404	0.89	0.869
sand B3	72	239	1.64	0.847
ilmenite	146	157	0.94	0.836

size distribution on the transition velocity a fraction of sand B1 was sieved and particles smaller than 140 µm and larger than 250 µm have been removed, called sand B1S. In addition to this, a mixture called sand B3 of 20 wt.-% sand A, 60 wt.-% sand B1 and 20 wt.-% sand B2 was produced to broaden the particle size distribution. Sand B1, sand B1S and sand B3 have similar values for the median diameter. But, as the difference of the particle size at 90 % and 10 % of the cumulative distribution divided by the median value show, the broadness of the distributions varies significantly. As a reference material for Geldart's group A the fraction sand A, being the only fraction not belonging to group B, was used for determination of transition velocities.

To investigate the influence of particle density on the fluid dynamic behavior, ilmenite was used having a slightly smaller mean particle size than sand B1.

The different fractions sand B1, sand B1S, sand B2, sand A and ilmenite have a monomodal particle size distribution each as it can be seen in the probability density distributions in figure 3.6 (a). In contrast to this, due to sand B3 being a mixture from other fractions it shows a clear bimodal distribution.

The determination of the sphericity Ψ of the materials shows that the particles are close to spherical shape. Due to edges in the particle shape they deviate from an ideal sphere as the values in table 3.1 show.

3.2.2 Density

Solid density, apparent density and bulk density are important parameters to characterize a bed material. The mean values of solid and bulk density, determined for the different bed materials used in this work, are listed in table 3.2.

Tab. 3.2: Bed material properties: densities, fixed bed concentrations and minimum fluidization velocities.

material	ρ_s $[\mathrm{kg\,m^{-3}}]$	ρ_b $[\mathrm{kg\,m^{-3}}]$	c_V $[-]$	U_{mf} $[\mathrm{m\,s^{-1}}]$	$U_{mf,calc}$ $[\mathrm{m\,s^{-1}}]$
sand A	2642	991	0.375	–	$1.5*10^{-4}$
sand B1	2599	1405	0.544	0.073	0.030
sand B1S	2599	1341	0.516	–	0.041
sand B2	2634	1551	0.589	0.140	0.101
sand B3	2614	1456	0.557	–	0.004
ilmenite	4465	2271	0.509	0.030	0.031

3.2.2.1 Solid Density

The solid density ρ_s is defined as the mass of solid material m_s divided by the volume of the solid material V_s of a sample:

$$\rho_s = \frac{m_s}{V_s} \tag{3.1}$$

It is measured using a MutliVolume Helium Pycnometer 1305 (Micromeritics Instrument Corporation). This pycnometer works with a gas pressure in the range of $140 - 170\,\mathrm{kPa}$. The sample size is $5\,\mathrm{cm^3}$ with a measurement accuracy of $0.07\,\mathrm{cm^3}$.

3.2.2.2 Apparent Density

In contrast to the solid density, the apparent density ρ_p, or particle density, also considers the pore volume of the particles V_p and is defined as given by

$$\rho_p = \frac{m_s}{V_s + V_p}. \tag{3.2}$$

If the bed material is porous, this density must be considered in fluid dynamics as characterizing particle property. Measurements of the apparent density of the different bed materials via liquid pycnometry of known volume ($50\,\mathrm{mL}$) showed that the particles have no open pores. Apparent density equals solid density and for this reason, it is not listed additionally in the table.

3.2.2.3 Bulk Density

In addition to the solids and pore volume, the bulk density does also consider the volume between particles V_b and is defined according to

$$\rho_b = \frac{m_s}{V_s + V_p + V_b}. \tag{3.3}$$

The bulk density of the different materials is determined according to the German standard DIN EN ISO 60 [171].

3.2.2.4 Fixed Bed Solid Concentration

The solid concentration of a fixed bed is defined as the volume of particles divided by the volume of the bulk. Thus, it can be determined from bulk density and apparent density and if the particles do not have pores, from bulk density and solid density as given by

$$c_V = \frac{V_s}{V_s + V_b} = \frac{\rho_b}{\rho_s}. \tag{3.4}$$

This parameter is needed for solids concentration measurements as calibration and reference value. The fixed bed solids concentrations for the different bed materials are given in table 3.2.

3.2.3 Minimum Fluidization Velocity

The minimum fluidization velocity of the bed material was measured in the laboratory fluidized bed facility FB100 having a diameter of 0.1 m (section 3.1.2). The static bed height was set to 10 cm and the pressure drop between windbox and 5 cm below the diameter extension was measured with decreasing superficial gas velocity from a clearly bubbling state down to fixed bed. The pressure drop of the porous gas distributor plate in empty operation was subtracted from the measured pressure drop and the minimum fluidization velocity was determined at the point where the increasing pressure drop in the fixed bed state changes to a constant pressure drop in the bubbling state. Measurements are repeated at least two times with changing the fraction of material used. The mean values measured can be found in table 3.2. Additionally, the minimum fluidization velocity is calculated using the approach of equation 2.2. The results are also listed in table 3.2.

3.2.4 Terminal Velocity

The terminal velocity of a single particle U_t can be calculated from a force balance of buoyancy force, gravitational force and drag force, which results in

$$U_t = \sqrt{\frac{4}{3} \frac{1}{c_{D,p}} \frac{(\rho_p - \rho_f)}{\rho_f} g \, d_p} \, . \tag{3.5}$$

It depends on the drag coefficient of the particle $c_{D,p}$, the particle density ρ_p, the fluid density ρ_f, the particle diameter d_p and the gravitational acceleration g. For spherical particles different correlations to calculate the drag coefficient are available. A summary is given by Cheng [172] with a correlation proposal:

$$c_{D,p} = \frac{24}{Re_{p,t}} \left(1 + 0.27 Re_{p,t}\right)^{0.43} + 0.47 \left(1 - exp\left(-0.04 Re_{p,t}^{0.38}\right)\right) \tag{3.6}$$

The drag coefficient of a single particle depends on the particle Reynolds number $Re_{p,t}$ at terminal velocity, the particle size, fluid density and fluid viscosity η_f:

$$Re_{p,t} = \frac{U_t d_p \rho_f}{\eta_f} \tag{3.7}$$

Using equations 3.5, 3.6 and 3.7 the terminal velocity distributions of the different bed materials can be calculated iteratively. As can be seen in figure 3.7, the probability density terminal velocity distribution of sand B1 and ilmenite are very similar. This can be related to the fact that the particle size distributions of both materials differ from each other, but the density of ilmenite is much larger compensating the difference in size. This might lead to similar fluidization behavior. In contrast to this, the terminal velocity of sand B2 is larger in average.

3.3 Measurement Techniques and Evaluation Methods

To investigate the fluid dynamic behavior in the fluidized beds, pressures are measured and evaluated. In addition to this, capacitance probes are used for detailed local flow analysis and entrainment is measured in the standpipe of the CFB400. The procedures and evaluation methods of these different measurement techniques are described in the following.

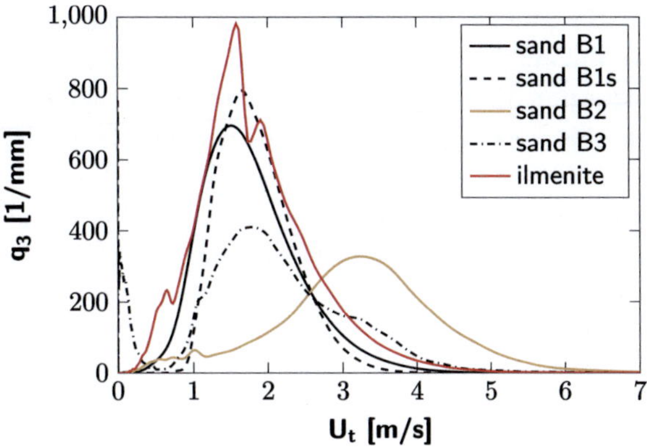

Fig. 3.7: Terminal velocities calculated from the probability density distributions of the bed materials used in this work.

3.3.1 Pressure Measurements

Pressure data has been recorded to get an estimation of the solids distribution and to determine pressure fluctuations. To get information about the axial solids concentration distribution from pressure data differential pressures have been recorded along the height of the fluidized bed in the three larger plants CFB100, CFB400 and FB1000. The data was recorded during steady state operation. If a parameter like the superficial gas velocity was varied, the system was given a time of at least 60 s to reach steady state. The mean value was calculated from data points of at least 30 s recording at steady state. The time-, radial- and height segment averaged mean solids concentration was approached using equation 2.5.

In addition to this, the pressure loop was calculated for the fluidized bed plants with solids return. The pressures relative to the ambient at each point of differential pressure measurement have been calculated from a point of relative pressure measurement and adding or subtracting (depending on position) the differential pressures in between. A pressure profile with a continuously decreasing pressure at increasing height is the result.

To analyze pressure fluctuations the standard deviations of the different pressure measurements (differential and relative to the ambient) are calculated. The results are plotted vs. the superficial gas velocity according to Bi *et al.* [6] as shown in figure 2.6. For the determination of the location of the maximum of this curve the data points are filtered using the Savitzky-Golay filter [173]. As described in section 2.3.2, change of measurement position and the evaluation of absolute or differential pressure data can lead to different transition velocities.

For this reason, the pressure measurements in the windbox and the freeboard, both relative to the ambient (comparable to absolute pressure measurements), are taken for determination of the transition velocity to compare the plants of different sizes in this work. Due to the limited maximum superficial gas velocity the turbulent regime was not achieved in the FB1000.

In the fluidized bed plants with a diameter up to 0.1 m gas is supplied by a pressurized air network. In the CFB400 two different roots blowers are used, having different piping systems of different volumes, respectively. The piping system of the smaller air blower has a volume of $0.28\,\mathrm{m^3}$, whereas the larger one has a volume of $0.71\,\mathrm{m^3}$ till the air distributor. The different volumes in the piping systems result in a difference of the pressure fluctuation intensities. Thus, they influence the standard deviation of the pressure in the windbox making the determination of the transition from bubbling to turbulent fluidization insufficient. Due to this effect, in the CFB400 pressures measured in the freeboard are taken for determination of the transition velocities only.

Additionally, frequency analysis of the pressure data was done, which is a common tool for determination of the pressure characteristics in fluidized beds [174–177]. The power spectral density function $S_P(\omega)$ is usually defined as the Fourier transform of the autocorrelation function $R_P(\tau)$ of the stationary pressure signal $P(t)$ as given by equation 3.8 [178]:

$$S_P(\omega) = \int_{-\infty}^{\infty} R_P(\tau)exp(-j2\pi\omega\tau)d\tau \tag{3.8}$$

Here, ω is the frequency and τ the time step the autocorrelation depends. A distribution of frequencies is the result. The probability density function of a scaled normal distribution (normal distribution [179] with added scaling factor s) as given by equation 3.9 was fitted to the frequency spectra to determine the arithmetic mean frequency μ, its arithmetic standard deviation σ and its scaling factor s:

$$f(x) = \frac{s}{\sqrt{2\pi}\,\sigma}exp\left(-0.5\left(\frac{x-\mu}{\sigma}\right)^2\right) \tag{3.9}$$

3.3.2 Capacitance Probe Measurements

To determine local flow parameters such as solids flow or bubble properties, different measurement techniques are used and described in literature, such as e.g. photography (surface or in quasi two-dimensional fluidized beds), heat transfer probes, suction probes for isokinetic solids sampling, capacitance probes, fiber optical probes, electro-resistive probes, tomography (X-ray, γ-ray, electric capacitance (ECT)) or magnetic resonance imaging (MRI) [13, 84, 180]. Each measurement technique has advantages and disadvantages in comparison to the others.

Capacitance probes have been found to be a reliable tool for determination of the local solids hold-up and bubble properties [99, 102, 110, 114, 181–186]. A major advantage of this measurement technique is its applicability to measurements at combustion conditions in high concentration and dilute flows [183, 184, 187].

3.3.2.1 Measurement Setup and Calibration

The basic working principle of a capacitance probe is the change of the dielectric constant in the electric field between two electrodes when the solids concentration changes. In a nutshell, a capacitance probe is a capacitor with a changing dielectric. The change of the dielectric constant can be related to the solids concentration. Different approaches have been introduced in literature to calculate the solids concentration from the changing dielectric constant, which is measured by the probe as summarized by Wiesendorf and Werther [184]. According to them, the linear relationship given in equation 3.10 fits solids concentrations in gas-solid systems well at ambient conditions:

$$c_V = c_{V,fb} \frac{K_e - K_{e,f}}{K_{e,fb} - K_{e,f}} \tag{3.10}$$

In addition, they introduced a relationship given in equation 3.11 which is applicable at high temperature conditions, because the dielectric constant of a gas-solid suspension was found to be highly dependent on temperature:

$$c_V = c_{V,fh} \frac{K_e - K_{e,f}}{K_{e,fb} - K_{e,f}\left(1 + \frac{\left(K_{e,fb}-K_e\right)\left(K_{e,fb}-\beta\right)}{K_{e,fb}K_{e,f}}\right)} \tag{3.11}$$

In both equations c_V is the solids concentration, $c_{V,fb}$ the fixed bed solids concentration, K_e the dielectric constant of the suspension, $K_{e,fb}$ the dielectric constant at fixed bed concentration and $K_{e,f}$ the dielectric constant of the fluidizing fluid. The parameter β is a fitting parameter, which was found to have a value of 3 according to Wiesendorf and Werther [184].

In this work the same measurement setup was used as by Wiesendorf and Werther [184]. The capacitance probe consists of two channels containing three electrodes each as shown in figure 3.8. A core electrode made from wolfram is surrounded by a guard electrode. Ceramic is used as insulation in between. In the amplifier system used (capaNCDT Series 600 by Micro-Epsilon Messtechnik GmbH & Co. KG), an alternating voltage magnitude is set to the core electrode.

To protect the measurement from other influences and guarantee a measurement only at the probe tip the same voltage is set to the guard electrode using an operational amplifier. The guard electrode is surrounded by a ground electrode. The measurement finally takes place in the electric field between the electrodes in the volume where the core electrodes protrude the probe body as shown in the figure. The AC sensor voltage is amplified and rectified to a DC signal voltage. A reference voltage $U_{V,ref}$ of 10 V is set against the signal voltage leading to a signal inversion. According to Wiesendorf and Werther [184] the dielectric constants measured are then calculated as given:

$$K_e = \frac{U_{V,ref} - U_{V,0}}{U_{V,ref} - U_V} \qquad (3.12)$$

Thereby, $U_{V,0}$ is the base capacitance voltage referring to the base capacitance in vacuum and U_V is the measured voltage. Further detailed information about the setup can be found in literature [184, 187].

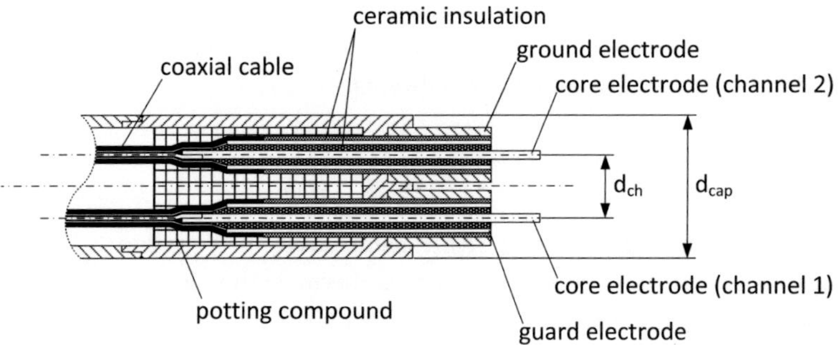

Fig. 3.8: Schematic of a capacitance probe as used in this work and [123].

To calculate the solids concentration, in this work a simplified approach proposed by Hage and Werther [183], as given in equation 3.13, was used:

$$c_V = c_{V,fb} \frac{U_V - U_{V,f}}{U_{V,fb} - U_{V,f}} \qquad (3.13)$$

In this approach the solids concentration is directly calculated from the output voltages of the probe U_V (suspension), $U_{V,f}$ (fluid) and $U_{V,fb}$ (fixed bed). According to Wiesendorf and Werther [184] this approach has not the same physical meaning as the realtion by the dielectric constants but was found to deliver appropriate results for the solids concentration comparable to equation 3.10.

Capacitance probe measurements have been carried out in the plants FB100, CFB400 and FB1000. Probes of different sizes are used in the facilities to withstand the fluid dynamic forces but to keep the probes small enough to mitigate the influence of the probe itself on the flow behavior. Because in larger plants the forces are greater, three different probes were used with larger probe diameter d_{cap} in larger beds. With increasing probe size the distance between the two channels d_{ch} of each probe was also increased. The probes dimensions can be found in table 3.3. The measurement positions have been varied axially and radially along the fluidized bed. The axial positions are given by the probe ports installed as given in section 3.1 for the different facilities. Radial positions investigated are given by the dimensionless radius r/R in the table for each plant.

Tab. 3.3: Characteristic properties of the capacitance probes and their usage in the fluidized bed facilities.

Probe	Used in	d_{cap} [mm]	d_{ch} [mm]	r/R [−]
CP1	FB100	8	3.5	0.9, 0.8, 0.6, 0.4, 0.2, 0
CP2	CFB400	16	5.9	0.95, 0.9, 0.75, 0.5, 0.25, 0, -0.25, -0.5, -0.75, -0.9, -0.95
CP3	FB1000	22	6.6	0.96, 0.9, 0.8, 0.6, 0.4, 0.2, 0

The voltages of both channels are recorded at a frequency of 10 kHz. The measurement duration depends on the measurement position. At the fluidized bed wall, where no bubbles have been expected, the duration was 60 s to determine the solids concentration distribution only. In cases where bubbles rise a measurement duration of 600 s was chosen to measure a representative distribution of the bubbles.

For measurements of the solids concentration the channel at the lower position (channel 1) is used, because, due to the upwards directed main flow direction, the upper channel lays in the wind shadow of the lower one influencing solids concentration results.

To determine the probability density function of the solids concentration from the histograms, kernel density estimation was used.

3.3.2.2 Determination of Bubble Properties

To detect bubbles and determine their properties a phase border between bubble and suspension phase must be defined. Measurement of the transition between both phases can be influenced due to different reasons. The capacitance probe has a certain measurement volume according to the electric field induced by the electrodes. If the probe does not penetrate the bubble with the whole measurement volume, both phases are measured at the same time. The result is an average concentration determined by the capacitance probe. In addition to this, contrary to the two-phase model, which assumes the bubble phase to be solids-free, the bubble phase may contain considerable amounts of solids. The transition between both phases is not sharp as it is the case in gas-liquid systems. Furthermore, a capacitance probe is an invasive measurement technique influencing the flow behavior slightly. Particles are decelerated by the probe or stuck at it leading to an increased concentration in the bubble and suspension phase. In the bubble phase, these issues can lead to the measurement of concentrations close to bubble phase concentration but not exactly to a solids concentration of zero. For this reason, the bubble phase is represented by a range of solids concentrations close to a value of zero and the phase border must be defined accordingly. The definition of the border and its derivation is explained in detail in the evaluation of the measurements in section 5.1.3.

Using a definition for the bubble-suspension phase border the detection of the presence of bubbles follows a specific algorithm. Nearby or incomplete passes of bubbles at the probe measurement volume can lead to short time signal crossing of the phase border. In this case the signal information is not sufficient to determine bubble properties. In addition to this, electrostatic charging in larger facilities can lead to spontaneous discharges at the probe tip. This is visible in the voltage signal recorded and can also lead to an apparent short time crossing of the phase border. Due to these effects a minimum time for border passing was set to $5\,\mathrm{ms}$. Passes of a duration below this time are not considered to be bubbles. This means for bubbles rising with a velocity of $2\,\mathrm{m\,s^{-1}}$ that a minimum bubble size is set to a length of $1\,\mathrm{cm}$. Bubbles having such a velocity and size are improbable to exist because they usually need to be larger to travel at these velocities [13].

The short time crossing described above, especially the one induced by electrostatic charging, can also happen during the measurement of a bubble. For this reason, to distinguish between two bubbles and to be sure that the signal was not disturbed during measurement of one bubble, a minimum duration between two bubbles of $2\,\mathrm{ms}$ was defined. This means bubbles detected close to each other occurring at a temporal distance below this value are counted as one bubble. Thus, for bubble rising at a velocity of $2\,\mathrm{m\,s^{-1}}$ the threshold of the spatial distance to each other is $4\,\mathrm{mm}$. Two bubbles rising at such a small distance to each other is unlikely to happen unless coalescence processes already started.

As mentioned before, all capacitance probes used in this work have two channels. These channels are aligned vertically to each other during measurement with one being in the flow shadow of the other as shown in figure 3.9. If a bubble passes the capacitance probe during its rise the bubble first enters the measurement volume of the lower channel before it further rises to reach the upper channel (cf. figure 3.9 (a)-(c)). The result is a time lag between both channel signals when significant flow events such as bubble rise occur. To be sure to measure the same bubble at both channels an overlap check of the detected bubbles in both signals is done. Thus, only bubbles having a minimum length like the distance between both channels can be counted. All bubbles fulfilling the conditions of the algorithm are counted. The amount of bubbles detected is related to the length of the whole signal to determine the local bubble rise frequency ω_B. To determine the rise velocity of a single bubble U_B the time lag τ between both signals must be determined. A popular method for determination of this lag is the cross-correlation method [132]. In this work it is determined using the cross-covariance function $\psi_{c_{V1}c_{V2}}$ given by

$$\psi_{c_{V1}c_{V2}} = \lim_{t_{max}\to\infty} \frac{1}{2t_{max}\sigma_{c_{V1}}\sigma_{c_{V2}}} \int_{-t_{max}}^{t_{max}} \left(c_{V1}(t) - \overline{c_{V1}}\right)\left(c_{V2}(t+\tau) - \overline{c_{V2}}\right) dt,$$

$$(3.14)$$

which directly yields the Pearson correlation coefficient R [188] of both signals for each time step. In this function $\sigma_{c_{V1}}$ and $\sigma_{c_{V2}}$ are the standard deviations and $\overline{c_{V1}}$ and $\overline{c_{V2}}$ the mean values of the time t dependent solids concentrations c_{V1} and c_{V2}, respectively. t_{max} is the maximum time of integration.

The advantage of using the cross-covariance function is the direct determination of the correlation coefficient which allows signal comparison simultaneously to the lag determination. The correlation coefficients can have values between -1 and 1 with the highest being identical signals. Thus, at the maximum of the cross-covariance function both signals have their highest similarity. To determine the time lag of the signals this maximum was determined for signals with a correlation coefficient above 0.9 at this point. This value was chosen to guarantee a high similarity in the signals.

The time lag τ determined (cf. figure 3.9 (c)) and the distance between both channels d_{ch} (which is known) lead to the velocity of a bubble U_B:

$$U_B = \frac{d_{ch}}{\tau} \qquad (3.15)$$

By determination of the time a bubble needs to pass one channel t_B (cf. figure 3.9 (d)) multiplied by the bubble velocity the pierced length l_p (cf. figure 3.9 (e)) can then be determined:

$$l_p = U_B t_B \qquad (3.16)$$

In evaluation of the pierced lengths of bubbles it must be kept in mind, that the resulting distribution of lengths is a result of the overlapping distributions

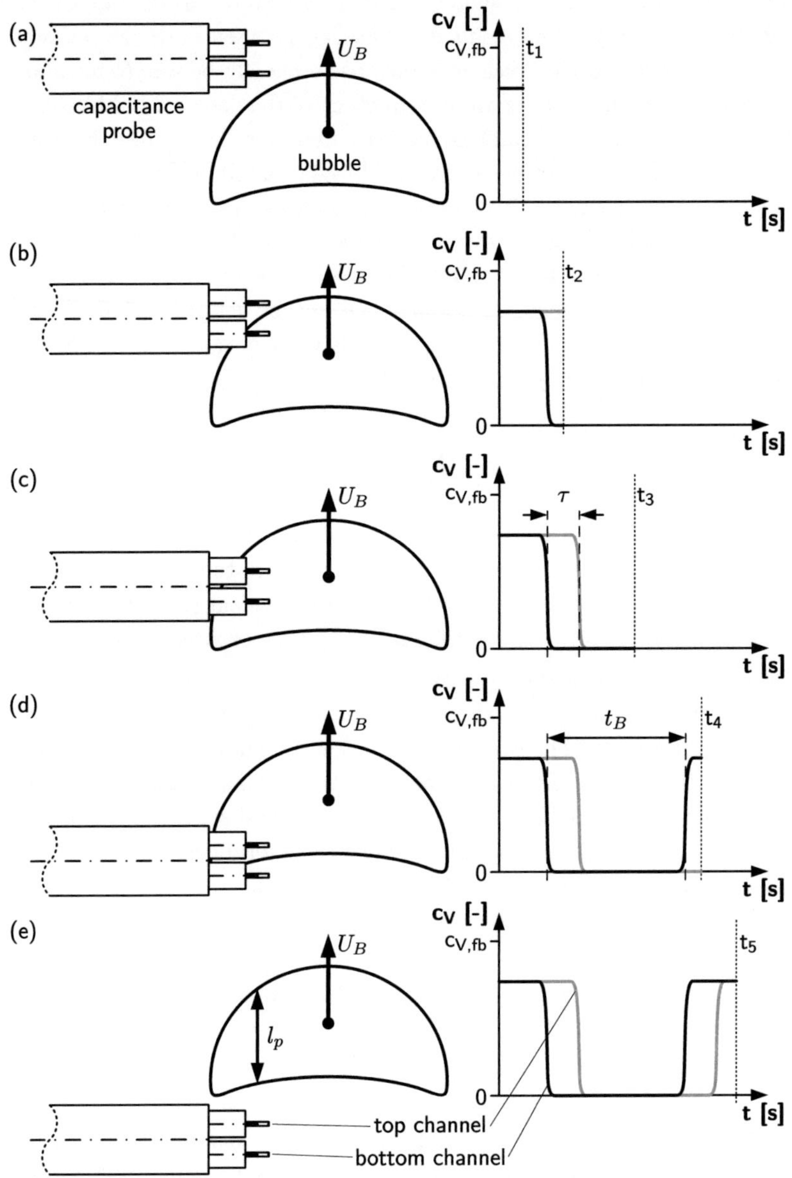

Fig. 3.9: A bubble passing a capacitance probe during its rise with the corresponding values of the solids concentration measured: (a) the bubble approaches the probe, (b) bottom channel enters the bubble, (c) both channels are inside the bubble, (d) bottom channel leaves the bubble, and (e) both channels have left the bubble.

of bubble size and position of bubble penetration. The pierced length measured depends on the point where the probe enters and leaves the bubble especially in the cases of spherical or cap shaped bubbles. For this reason, the pierced length underlies a statistic variation [64]. Superimposed to this are bubble interactions leading to a natural size distribution of bubbles.

Nevertheless, the distribution of bubble size as well as of bubble velocities can be described by a logarithmic normal distribution as it is the case in e.g. liquid-liquid systems [189]. The probability density function $f_X(x)$ of the log-normal distribution of parameter x is defined in equation 3.17 as given by Forbes *et al.* [190]:

$$f_X(x) = \frac{1}{\sqrt{2\pi}\,\sigma x} exp\left(-\frac{(ln(x) - \mu)^2}{2\sigma^2}\right) \tag{3.17}$$

Thereby, σ describes the standard deviation of the parameter's natural logarithm and μ its mean value. The arithmetic mean value $E(X)$ and the arithmetic standard deviation $S(X)$ of the function $f_X(x)$ are given by equations 3.18 and 3.19:

$$E(X) = exp\left(\mu + \frac{1}{2}\sigma^2\right), \tag{3.18}$$

$$S(X) = exp\left(\mu + \frac{1}{2}\sigma^2\right)\sqrt{exp(\sigma^2) - 1} \tag{3.19}$$

To compare pierced lengths and bubble velocities measured at different conditions, the arithmetic mean values of the distributed parameters are compared in this work.

3.3.3 Solids Circulation Rate

The solids circulation rate was estimated in the CFB400 only. As it can be seen in figure 3.3, a separately mounted section is installed in the standpipe of the facility. This section is connected via flexible segments to the standpipe. It consists of a butterfly valve at the bottom of the section with a 1 m long pipe having a diameter of 0.2 m above. A detailed sketch of this section can be found in figure 3.10. The solids circulation rate is determined by temporal closing of the butterfly valve to accumulate entrained solids above. By knowing the change of the mass Δm in the segment related to the time span Δt, the mass flow of solid material $\dot{m}$ can be determined. The solids circulation rate G_s is then determined by relating the mass flow to the cross-sectional area of the fluidized bed A_0:

$$G_s = \frac{\dot{m}}{A_0} = \frac{\Delta m}{\Delta t A_0} \tag{3.20}$$

To guarantee that the pressure loop is disturbed as little as possible, the separate section is bypassed by a pressure equalization pipe.

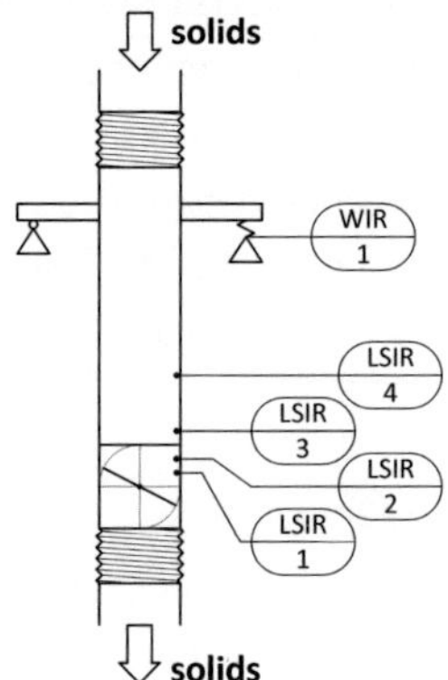

Fig. 3.10: Measurement system of the solids circulation rate installed in the CFB400 standpipe.

The measurement of the mass in the separate mounted section was done by two different methods. The section is mounted at two points, one equipped with a load cell to measure the change of weight. In addition to this, the level of solids was measured using four PTC-thermistors. The thermistors have been arranged at distances of 5 cm (LSIR1), 10 cm (LSIR2), 20 cm (LSIR3) and 40 cm (LSIR4) from the butterfly valve at the wall of the pipe as shown in figure 3.10. When the solid level reaches the hot sensors they cool down resulting in a change of their resistance. Thus, it is possible to measure the duration for the solid level to reach the different thermistors. By knowing the section volume between the thermistors and the bulk density of the solid material, the mass in the section can be determined.

Both types of measurements showed sufficient agreement to each other, whereas slightly higher solids circulation rates have been determined by the level measurement.

For each steady state five measurements of the solids circulation rate were done. The reproducibility was found to be high. The duration of closing the butterfly valve was chosen according to the amount of solids circulating. At large superficial gas velocities ($> 3\,\mathrm{m\,s^{-1}}$) large amounts of solids circulate reaching the level of the fourth PTC-thermistor (LSIR4) fast, whereas at low velocities the butterfly valve was closed partially for more than 30 s to reach the level of the third PTC-thermistor (LSIR3). After opening the butterfly valve a time of minumum 60 s was waited until the next measurement to guarantee steady state in the system.

4 Regime Identification

To analyze the flow structure of turbulent fluidized beds the flow regime must be identified. This is usually done by analysis of pressure fluctuations, which are in the center of interest in this chapter for this reason. The standard deviation of the pressure signals is evaluated to determine the regime transition velocities. Thereby, the fluidization conditions are varied and influences of different parameters are investigated. Finally, frequency analysis of the pressure signals is done. This chapter is partially based on results published by the author previously in [112] and [168].

4.1 Pressure Fluctuations in Fluidized Beds

In fluidized beds differential pressures (between two points in the bed) and absolute pressures (single point) can be measured. The analysis of these two type of pressures for determination of the transition velocity from bubbling to turbulent fluidization can lead to different results [6, 119]. Differential pressures are measured between two different heights above the gas distributor. Thus, the differential pressure signals are the difference of the absolute pressures measured at each point. Absolute pressures are influenced by the local flow structure which is e.g. in bubbling beds characterized by the rise of bubbles. Due to their pressure profiles (see figure 2.3), the rise of multiple bubbles having a size and velocity distribution induces fluctuations in the local and global absolute pressure. Thus, differential pressures measured are the result of height dependent changes of the flow structure combined with the time shifted occurrence of distinctive flow events due to the spatial distance of the pressure ports to each other. This enhances the difficulty in interpretation of differential pressure signals for analysis of the local flow structure in comparison to absolute pressure measurements. For this reason, absolute pressure measurements are preferred in this work to obtain conclusions on the flow behavior.

In all fluidized bed facilities used in this work pressures relative to the ambient are measured in the windbox, in the fluidized bed and in the freeboard. If the ambient pressure is constant during the experiment, the pressure fluctuations measured equal the pressure fluctuations of an absolute pressure measurement.

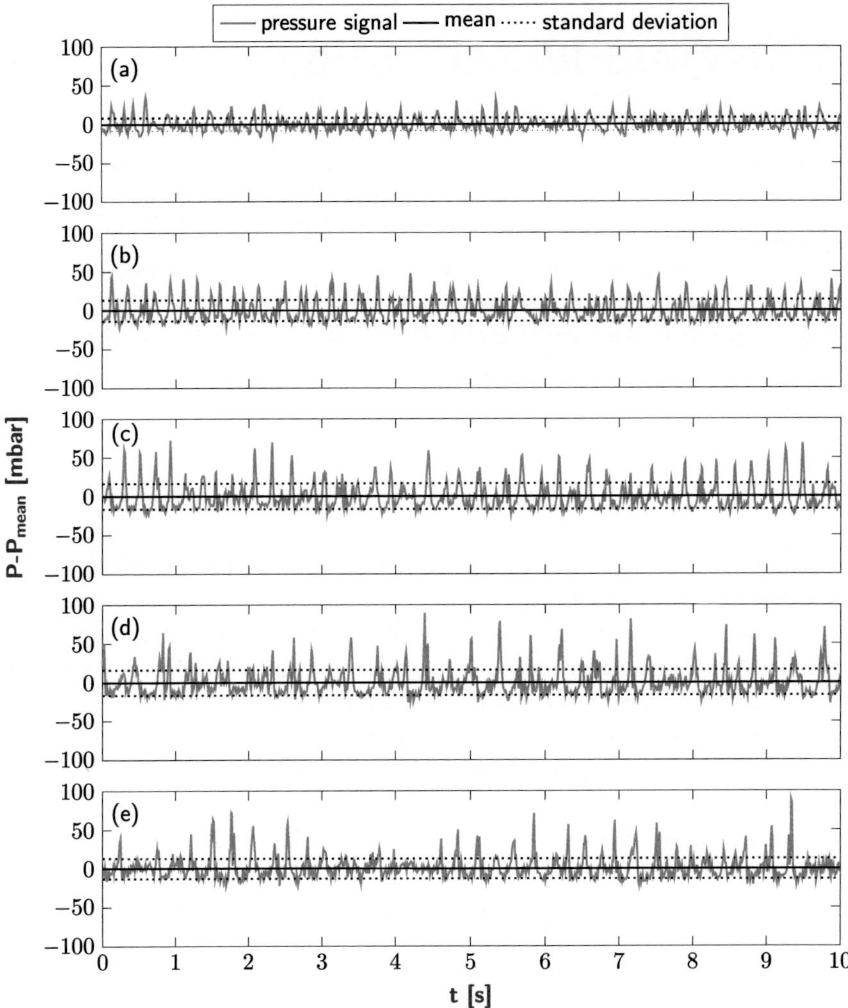

Fig. 4.1: Normalized pressure signals (measured relative to the ambient in the windbox), their mean values and standard deviations recorded for ilmenite at $D = 0.1\,\text{m}$ and $H_0 = 0.15\,\text{m}$ at different superficial gas velocities having values of (a) $U_0 = 0.25\,\text{m}\,\text{s}^{-1}$, (b) $U_0 = 0.51\,\text{m}\,\text{s}^{-1}$, (c) $U_0 = 0.77\,\text{m}\,\text{s}^{-1}$, (d) $U_0 = 1.2\,\text{m}\,\text{s}^{-1}$ and (e) $U_0 = 1.58\,\text{m}\,\text{s}^{-1}$.

Exemplary sections of pressure signals subtracted by their mean values at different superficial gas velocities are shown in figure 4.1. In the clear bubbling regime in figure 4.1 (a) the maximum amplitude of pressure fluctuations is small in comparison to larger superficial gas velocities. Irregular, random peaks can be observed having different amplitudes. In contrast to this, by visual observation of (b) and (c) the magnitude of the amplitudes increase and the peaks tend to occur more regular in similar temporal distance to each other. The larger amplitudes lead to an increase in the standard deviation of the pressure signal. If it is assumed that larger bubbles induce larger pressure fluctuations, which is plausible due to their larger pressure gradients (see figure 2.3), it can be concluded from the pressure profiles that larger bubbles rise at larger superficial gas velocities. Literature generally agrees to this for the bubbling fluidized bed regime [13]. Due to coalescence phenomena and bubble formation processes bubbles always occur in a statistic distribution. Thus, the amplitudes of the pressure fluctuations also occur in a statistic distribution which explains the non-uniformities of the amplitudes that can be seen in figure 4.1. By further increase of the superficial gas velocity in (d), larger maximum amplitudes can be found than in (c). The occurrence of these large peaks gets more irregular with peaks of different sizes in between. Because the pressure fluctuation amplitudes decrease in their mean, the standard deviation of the pressure signal decreases. According to the definition of the transition from bubbling to turbulent fluidization, the turbulent regime must be reached in (d), whereas (c) marks the region of transition. If the bubble size is linked to the pressure fluctuations a broad range of bubble sizes must be the result in the turbulent regime. Maximum bubble sizes grow further in the turbulent regime as the maximum amplitudes increase. This statement is supported by local probe measurements in turbulent beds in some studies [108, 110]. In figure 4.1 (e) the turbulent flow regime should be well developed. The standard deviation of the pressure further decreases. Large pressure amplitudes occur rarely and pressure fluctuations are still irregular.

The probability density distributions of the pressure signals in the plots (a), (c), (d) and (e) in figure 4.1 are shown in figure 4.2, respectively. From the state of bubbling fluidization in figure 4.2 (a) to the point of transition to turbulent fluidization in (b) the distribution clearly broadens. Whereas in (a) the pressure fluctuates in a range of 60 mbar, in (b) the range almost doubles in size. The distribution functions can be described by logarithmic normal distributions. The standard deviations (std) shown in the figure are calculated under the assumption of the pressure having a normal distribution which is usually done in literature. Thus, the values do not give the actual arithmetic standard deviation of the distribution function which is defined according to equation 3.19. This must be considered in analysis of pressure signals for detailed flow structure.

Nevertheless, the calculation of the standard deviation defined for a normal distribution does not influence the determination of a qualitative conclusion about pressure fluctuation intensities.

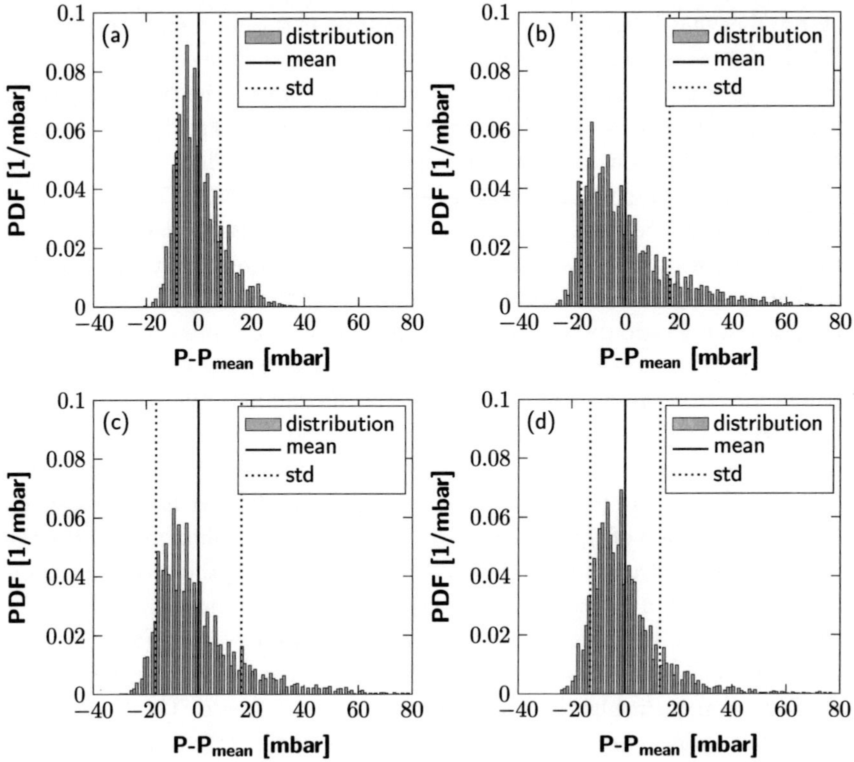

Fig. 4.2: Probability density distributions of normalized pressure signals (measured relative to the ambient in the windbox), their mean values and standard deviations recorded for ilmenite at $D = 0.1\,\text{m}$ and $H_0 = 0.15\,\text{m}$ at different superficial gas velocities having values of (a) $U_0 = 0.25\,\text{m\,s}^{-1}$, (b) $U_0 = 0.77\,\text{m\,s}^{-1}$, (c) $U_0 = 1.2\,\text{m\,s}^{-1}$ and (d) $U_0 = 1.58\,\text{m\,s}^{-1}$.

In figure 4.2 (c) and (d) the range of pressure fluctuations decreases with increasing superficial gas velocity. This results in a decrease of the standard deviations. To determine the transition from bubbling to turbulent fluidization, as well as the transition from turbulent to fast fluidization, it is common to plot the standard deviation of the pressure in dependence of the superficial gas velocity [6]. This is shown exemplary in figure 4.3 (including the values from the figures 4.1 and 4.2). A maximum of pressure fluctuations and thus the point of transition from bubbling to turbulent fluidization can be clearly observed.

Depending on bed material properties the peak in the standard deviation can stretch over a broad range of superficial gas velocities as it is the case in the figure. For this reason, it can be assumed that the transition from bubbling to turbulent fluidization occurs smoothly indicating a transient transition of the flow structure.

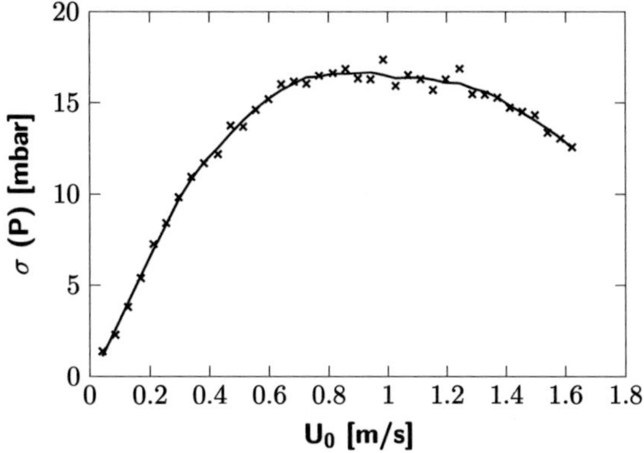

Fig. 4.3: Standard deviation of the pressure measured in the windbox for ilmenite at $D = 0.1\,\text{m}$ and $H_0 = 0.15\,\text{m}$.

4.2 Transition from Bubbling to Turbulent Fluidization

The transition from bubbling to turbulent fluidization is influenced by variation of several parameters, such as particle properties, gas properties and fluidized bed dimensions. In addition, change of measurement location and method can lead to different results as discussed before. In literature no consensus exists about the measurement location of transition velocities using the method of standard deviation. Single point (absolute) and two point (differential) measurements of pressure can deliver different results for the transition velocity and the transition is known to first occur at the top of the fluidized bed and to migrate downwards to the gas distributor with increasing superficial gas velocity [6, 119, 120]. For this reason, in this work measurements of the pressure fluctuations in the windbox of each facility and in the freeboard region, both relative to the ambient, are compared to guarantee no influences by the measurement position on the transition velocity.

4.2.1 Influences of Particle Properties

The kind of bed material used has large impact on the transition velocity. Main influencing parameters are particle size and distribution and particle density. Furthermore, it can be assumed that the shape of particles has an influence on the transition velocity from bubbling to turbulent fluidization because it influences fluidization behavior generally [191]. Particle size and density are considered in most correlations in literature because they are part of Reynolds and Archimedes number of which these correlations consist mostly (see section 2.3.2).

4.2.1.1 Particle Size and Distribution

Figure 4.4 shows the influence of particle size distributions on the transition for different fractions of sand under same conditions. In (a) the typical curves of the pressure standard deviation in dependence of the superficial gas velocity are shown for fractions having a different broadness of particle size distribution. The maximum values that the standard deviation reaches decrease when particle size distribution broadens. This means, the overall pressure fluctuation intensity is lower for beds having a broad particle size distribution. Under the assumption of the bubble size being linked to the pressure fluctuations it can be concluded that these beds have a lower mean bubble size. Reason for this can be larger inter-particle forces resulting from the larger amount of fines [72]. A clear statement for the location of the maximum cannot be made from the results due to the flat and broad peaks in the standard deviation. In literature the influence of the broadness of particle size distribution is described with larger transition velocities in beds having a narrow distribution [115, 121]. This effect is explained with a higher effective suspension phase viscosity in beds having a narrow distribution leading to a lower probability in splitting of bubbles.

In figure 4.4 (b) the results for the transition velocities of sand fractions having a different mean particle diameter is shown. Particles of the smallest fraction belong to group A according to Geldart's classification, whereas the others belong to group B. Beds with larger mean particle sizes have larger transition velocities. In these beds, particles have a higher inertia and larger forces are needed to reach similar bed expansion ratios than with small particles. Furthermore, more gas flows through the suspension phase of the bed as predictions of the minimum fluidization velocity show [10]. If the transition from bubbling to turbulent fluidization under similar conditions is assumed to happen at similar expansion ratios, larger transition velocities for larger particles are the result. Additionally, in beds of particles belonging to Geldart's group A inter-particle forces play a significant role influencing fluidization behavior and leading to smaller bubble

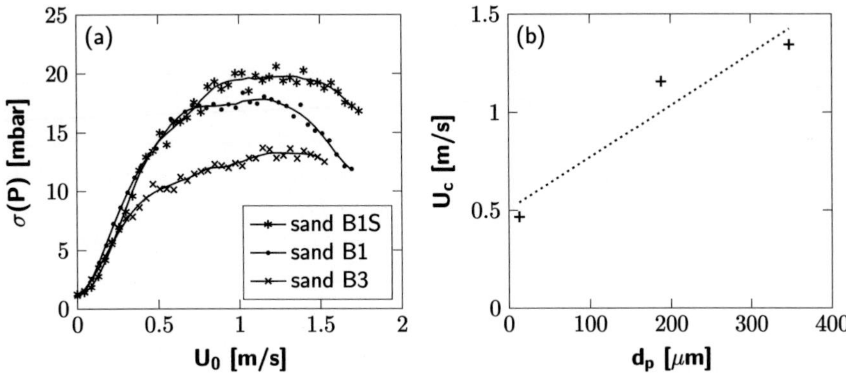

Fig. 4.4: (a) Influence of the broadness of the particle size distribution on the pressure standard deviation in dependence of the superficial gas velocity and (b) influence of mean particle size on the transition velocity U_c for different fractions of sand ($D = 0.1\,\text{m}$, $H_0 = 0.3\,\text{m}$) acc. to [168].

sizes as observed in literature [67]. Pressure fluctuation intensities in beds with larger particles are found to be larger than with smaller ones. This might be related to a larger bubble sizes occurring in larger beds.

4.2.1.2 Solid Density

In addition to particle size, the solid density is a major parameter for the categorization of fluidization behavior (see fig. 2.2). Fluid dynamics are influenced by this parameter and thus, the transition velocity, too. Larger densities at constant particle size lead to higher inertia and gravitational forces of the particles needed for regime transition at larger superficial gas velocities. Transition velocities of fractions of sand B1 and ilmenite are measured in this study both having similar terminal velocity distributions. Ilmenite, which has a larger solid density, has larger transition velocities than sand at same operating conditions as can be seen in figure 4.5.

4.2.2 Influences of Fluidized Bed Plant Dimensions

The construction of the fluidized bed is known to influence the transition velocity from bubbling to turbulent fluidization. Thereby, bed diameter and static bed height are two characteristic parameters which have also been investigated in literature [6, 101, 104, 112–114].

4.2.2.1 Static Bed Height

Measurements in this work have been limited to aspect ratios (static bed height divided by bed diameter) between 1 and 4. Due to a low amplitude of the pressure fluctuations in shallow beds with an aspect ratio below 1, it was not possible to determine a transition velocity in these beds. This was especially the case for the beds having diameters of 0.1 m or below. If the aspect ratio was adjusted above the range, slugging occurred, which made a determination of the transition velocity from bubbling to turbulent fluidization impossible.

Results for the transition velocities of three different bed materials in dependence of the static bed height are shown in figure 4.5. The transition velocity generally gets larger if the static bed height is increased. In literature the same trends are reported [6, 101, 104, 112]. Sand B1 and ilmenite show similar behavior in the values and change of the transition velocities. This can be related to similar fluidization behavior resulting from the similarity in their terminal velocity distributions.

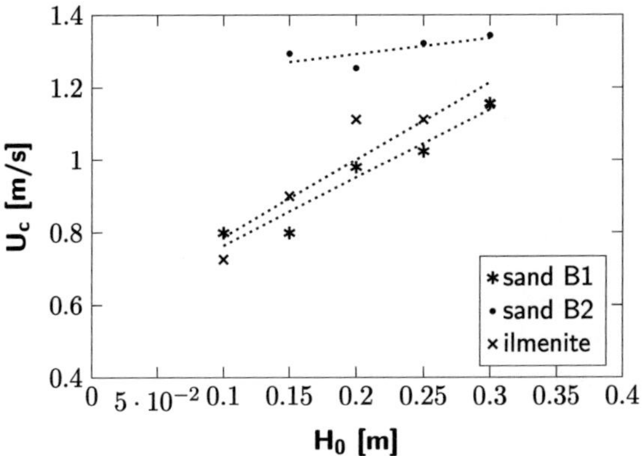

Fig. 4.5: Influence of static bed height and density on the transition velocity U_c for different materials ($D = 0.1\,\mathrm{m}$) acc. to [168].

The flow behavior in fluidized beds changes with distance from the gas distributor due to coalescence phenomena. Generally, larger bubbles can be found at larger heights [55–57, 59, 61, 64, 65, 68]. Due to their buoyancy these bubbles rise at larger velocities leading to a reduction of their local residence time. For this reason, in beds of larger static bed heights the mean bubble residence time is lower than in beds of a lower static bed height. This leads to a reduction of the bed expansion at same superficial gas velocities for larger bed heights. Thus, to reach similar void fractions in beds of different static bed heights, larger superficial gas velocities are needed in a bed of larger height. If the transition between the flow regimes is assumed to happen at similar void fractions, larger transition velocities are the result at larger static bed heights as it can be observed in figure 4.5.

In literature a variation of the transition velocities was found if measured at different distances from the gas distributor [6, 119, 120]. The regime transition was found to migrate downwards in the fluidized bed with increasing superficial gas velocity. This might be explained by reaching void fractions at larger heights earlier due to the high rate of particle ejection at the bed surface. To be able to compare plants of different sizes in this work, the pressure signals in the windbox and freeboard have been analyzed only. For this reason, such behavior could not be observed and an overall transition velocity was determined.

4.2.2.2 Bed Diameter

In the range of bed diameters investigated, the transition velocity from bubbling to turbulent fluidization is highly influenced by the diameter of a fluidized bed. This is shown in figure 4.6, where in (a) the dependence the transition velocity of the static bed height and in (b) of the aspect ratio are plotted for different diameters.

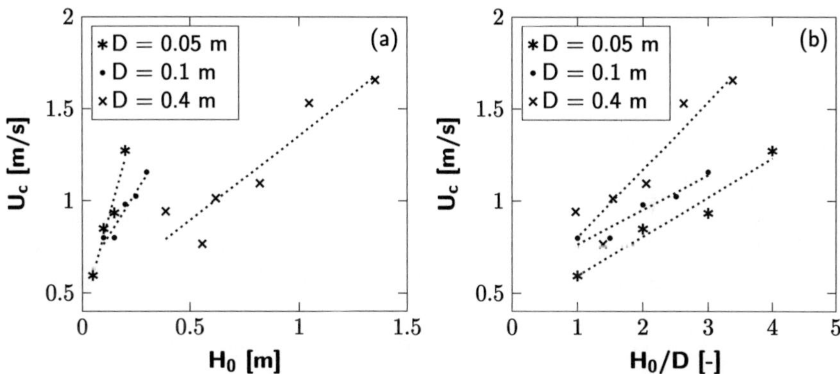

Fig. 4.6: Influence of bed diameter: transition velocity in dependence of (a) the static bed height and (b) the aspect ratio for sand B1 acc. to [168].

In the FB50 and the FB100 static bed heights have been adjusted being much lower than in the CFB400. Nevertheless, due the differences in bed diameter, the aspect ratios have been adjusted equally. Despite the difference in their static bed height, in all plants transition velocities in the same order of magnitude have been achieved. For this reason, it can be concluded, that in smaller beds larger transition velocities occur if the static bed height is the same in the facilities.

This can be related to the large wall influence in small fluidized bed plants. In these plants bubble growth is limited and bubbles quickly reach the size of the fluidized bed diameter. In contrast to this, in fluidized beds of larger sizes large regions of bubble rise are formed [55]. Bubbles rising in such regions are less exposed to the stress induced by the back flowing solids (mostly at the wall) than in beds where bubbles quickly reach the size of the diameter of the fluidized bed. The larger stress on bubbles at same superficial gas velocities but smaller bed diameters results in an earlier break-up of the flow structure. Thus, the flow structure is highly dominated by the bed diameter resulting in different transition velocities. The influence of the bed diameter on the transition velocity was found to be most intensive for beds smaller than 0.2 m in diameter according to Bi *et al.* [6].

The large size of bubbles in comparison to the bed diameter in small facilities causes a stronger interaction between rising bubbles and down flowing solids than in larger facilities. This leads to larger void fractions in smaller plants and a resulting larger bed expansion at same superficial gas velocities. In figure 4.6 (b) at equal aspect ratios lower transition velocities can be found for larger fluidized bed diameters. The result can be attributed to the differences in the expansion behavior.

4.2.3 Influence of Temperature

In addition to bed material and plant properties, the properties of the fluidization gas are known to influence the flow behavior in fluidized beds significantly. Gas properties in fluidization are mainly defined by gas density and viscosity. Both parameters are functions of the temperature with the viscosity increasing proportionally to T^γ ($0.6 \leq \gamma \leq 1$) and the density being inversely proportional to the temperature [192]. By change of the gas properties the drag acting on particles changes accordingly. Shabanian and Chaouki [72] describe an increase of the drag force on particles of Geldart's group A at higher temperatures as a result of the influence of the increasing viscosity. On the other hand, they describe a contrary effect for larger particles belonging to Geldart's group D where the decreasing density at larger temperatures leads to a reduction of the particle drag force.

Figure 4.7 (a) shows the standard deviation of the pressure in dependence of the superficial gas velocity at different temperatures. If the temperature is increased, the point of transition moves to larger superficial gas velocities and the maximum value of the standard deviation decreases. This behavior was found for all materials investigated as the dependency of the transition velocity of the temperature in figure 4.7 (b) shows. Larger transition velocities with increasing temperature are the result of the change in the drag acting on the particles. If the drag force for a single particle of the materials investigated in this work is calculated, it is found to increase with temperature. This leads to the conclusion that the change of viscosity still dominates the change of the flow structure for particles of Geldart's group B. The increasing particle drag force leads to a higher stability of the suspension phase of the fluidized bed. Larger amounts of gas are going through this phase at larger interstitial gas velocities. This is accompanied by a lower stability of bubbles as a result of the larger particle drag forces as reported in literature [72]. Moreover, bubbles decrease in size at a constant bubble passage frequency and bed expansion reduces [67, 72]. Thus, if the transition from bubbling to turbulent fluidization is assumed to happen at similar void fractions of the bed, larger superficial gas velocities are necessary at larger temperatures to reach these void fractions. An increase of the transition velocity at larger temperatures is the result as found in figure 4.7 and by other authors [72, 111, 113, 114, 116–118]. Furthermore, the reduction of the bubble size leads to a lower mean amplitude of pressure fluctuations and thus, standard deviation of the signal, which can be seen in figure 4.7 (a).

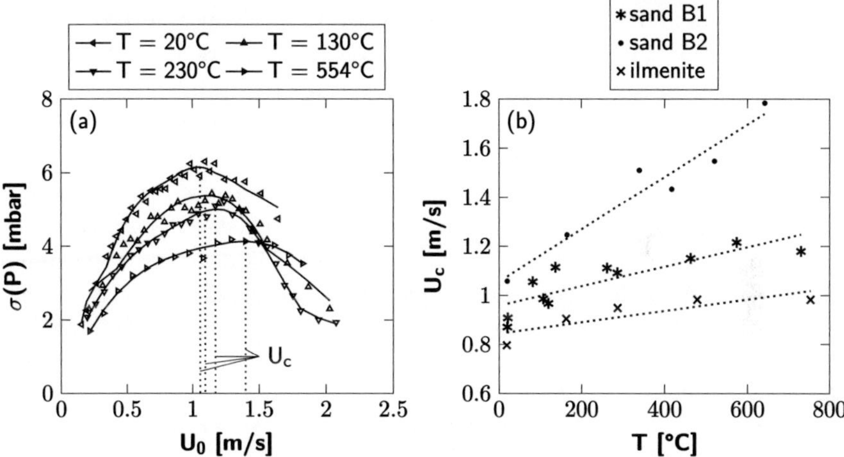

Fig. 4.7: (a) pressure standard deviation at different temperatures and (b) influence of the temperature on the transition velocity U_c ($D = 0.1\,\mathrm{m}$, $H_0 = 0.3\,\mathrm{m}$) acc. to [168].

The increase of the transition velocity of ilmenite and sand B1 in figure 4.7 (b) follows a similar slope. The terminal velocity distributions of both materials are similar. They are influenced equally by the change of temperature. For this reason, similar fluidization behavior can still be observed at larger temperatures and transition velocities change accordingly.

4.2.4 Correlative Approach

Several approaches to predict the transition velocity from bubbling to turbulent fluidization are introduced in literature (section 2.3.2). As explained above, several parameters influence the transition velocity. None of the correlations available in literature considers all parameters investigated in this work. For this reason, a new correlation is introduced by

$$Re_c = 0.2 Ar^{0.659} D^{0.293} + 2.146 \left(\frac{\eta_f \left(20°C \right)}{\eta_f \left(T \right)} \right)^{2.219} \frac{H_0}{D} \tag{4.1}$$

for a sufficient prediction of the influences of bed material properties, plant properties and gas properties. The approach describes the Reynolds number at the point of transition Re_c in dependence of the Archimedes number Ar, the bed diameter D, the gas dynamic viscosity η_f and the aspect ratio H_0/D. A linear behavior between the aspect ratio and the Reynolds number at transition is assumed from figure 4.6 (b) and was shown in a previous work [112].

In figure 4.8 the prediction of the transition velocity in dependence of the bed diameter of three different correlations from literature is shown. The results do not give an accurate prediction of the transition velocities measured for different aspect ratios. In addition, the predicted transition velocities of equation 4.1 are shown in the figure giving a good estimation of the trends measured.

The parity plot of the Reynolds numbers measured and predicted by equation 4.1 is given by figure 4.9. The results show a sufficient accuracy. The largest deviations have been found for high operating temperatures. Furthermore, it must be mentioned that for the investigation of the influences of gas properties only temperature was varied. The change of operating pressure or the kind of fluidizing medium does also lead to different gas densities and viscosities and thus, different fluidization behavior. Furthermore, to consider the broadness of particle size distribution in the correlative approach, further investigation is needed.

4.3 Transition from Turbulent to Fast Fluidization

Different approaches for the determination of the transition velocity from turbulent to fast fluidization are available in literature as summarized in section 2.3.7. In this work the regime transition is defined at the point where pressure fluctuations level off. Thus, the transition velocity U_k is determined.

Because the leveling off of the pressure fluctuations occurs gradually, the point of intersection of two straight lines, one fitted to the range of declining

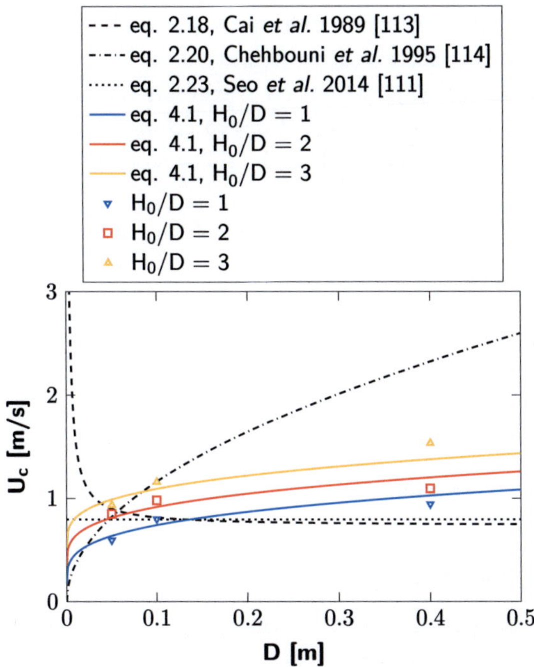

Fig. 4.8: Measured values of the diameter dependence of the transition velocity in comparison to correlations from literature and equation 4.1 for sand B1 at ambient conditions (cf. [168]).

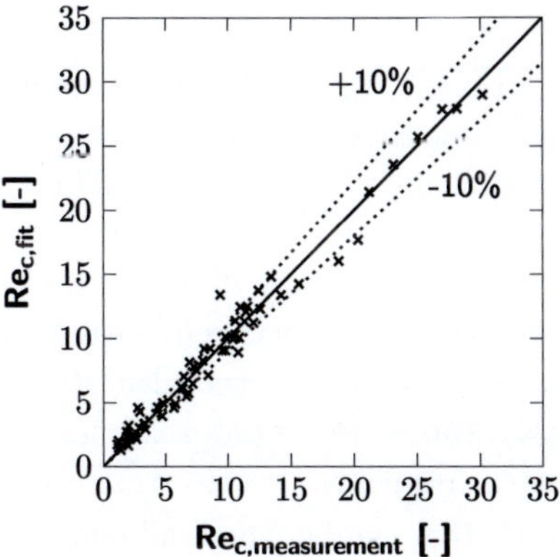

Fig. 4.9: Parity plot of the measured and calculated Reynolds number at point of transition from bubbling to turbulent fluidization.

pressure fluctuations (turbulent fluidization) and the other to the range of constant pressure fluctuations (fast fluidization) is determined. The superficial gas velocity at this point is defined as transition velocity U_k. This is shown in figure 4.10 (a) where the standard deviation of the pressure is plotted for superficial gas velocities reaching from bubbling to clear fast fluidization.

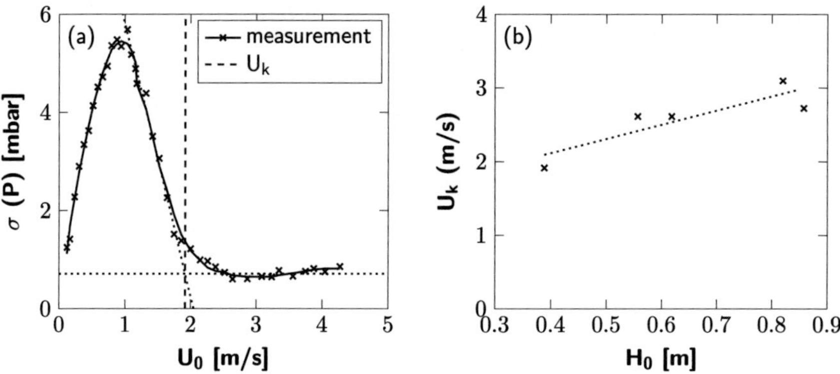

Fig. 4.10: (a) determination of the transition velocity U_k and (b) influence of static bed height on U_k ($D = 0.4\,\mathrm{m}$, sand B1).

Due to the design of the FB50, FB100 and FB1000 measurements of the transition velocity from turbulent to fast fluidization could not be carried out in these facilities. Measurements of U_k have only been carried out in the CFB400 at ambient conditions with sand B1 as bed material. Because fluidization behavior in the bubbling and turbulent regime are highly influenced by bed material properties, fluidized bed dimension and gas properties it can be assumed, that the transition between turbulent and fast fluidization is influenced by these parameters as well. Figure 4.10 (b) shows the influence of the static bed height on the transition velocity U_k. An increase of the transition velocity can be found in deeper beds. The same trend was found in literature [177, 193]. This is analog to a delayed transition from bubbling to turbulent fluidization. In beds having larger static bed heights a turbulent bottom zone can still exist, whereas in the top of the bed entrainment begins to become significant.

Most correlations for U_k that can be found in literature describe the Reynolds number at this point in dependence of the Archimedes number, as it is the case for U_c [6]. Thus, these correlations do not consider influences of some important process parameters as mentioned above. The introduction of a sufficient correlation for the prediction of the transition velocity from turbulent to fast fluidization needs further investigation.

4.4 Fluidized Bed Frequency

To get deeper understanding of the pressure fluctuation and regime transition behavior, frequency analysis of the pressure signal is a common tool in literature [174, 175, 177, 193–200]. For this reason, the power spectral density distributions of the pressure signals in the windbox at steady state are calculated for the different operation conditions investigated. In figure 4.11 these distributions are shown for ilmenite at different superficial gas velocities as exemplary case. The corresponding pressure signals can be found in figure 4.1. In figure 4.11 (a) at a low superficial gas velocity in the clear bubbling regime two major peaks can be found close to a frequency of 4 Hz. In beds of group B particles Kage *et al.* [174] found three principle frequencies which agree to bubble eruption frequency, bubble generation frequency and spontaneous frequency of the bed. The two major peaks shown in figure 4.11 (a) might be related to the frequencies found by Kage *et al.* [174]. If the superficial gas velocity is increased, a broad range of frequencies occurs as shown in (b)-(e). Bubble generation, coalescence and eruption are statistical processes with deviations leading to a spreading of frequencies. For this reason, the major frequencies described in literature could only be determined for some cases of sand B2, but not for sand B1 and ilmenite. The frequency analysis of the two latter materials lead to noisy power spectral density distributions reaching over a wide range of frequencies as they can be seen for ilmenite in figure 4.11.

The peaks of the power spectral density are found to increase from figure 4.11 (a) to (c) and decrease at even larger superficial gas velocities until (e). This is related to the regime transition occurring around the superficial gas velocity of (c).

The frequency analysis was done for the pressure measurements in the windbox of the fluidized beds. This guarantees comparability because the location of the pressure measurement is known to influence the pressure fluctuations and frequencies determined [177]. Significant influences on the frequency spectra in the windbox induced by the roots blowers in the larger plants as they are reported in literature [201] have not be found. This might be related to the size of the piping system in which large amounts of air are cushioning fluctuations induced by the air blower.

To compare the power spectral density distributions despite their spreading over a large range of frequencies, a scaled normal distribution (eq. 3.9) is fitted to them. These fitted curves can be found in figure 4.11 in addition to the power spectral density distributions. The mean value of the fitted distribution function equals the mean bed frequency. This frequency is shown in figure 4.12 (a) for three different bed materials in dependence of the superficial gas velocity.

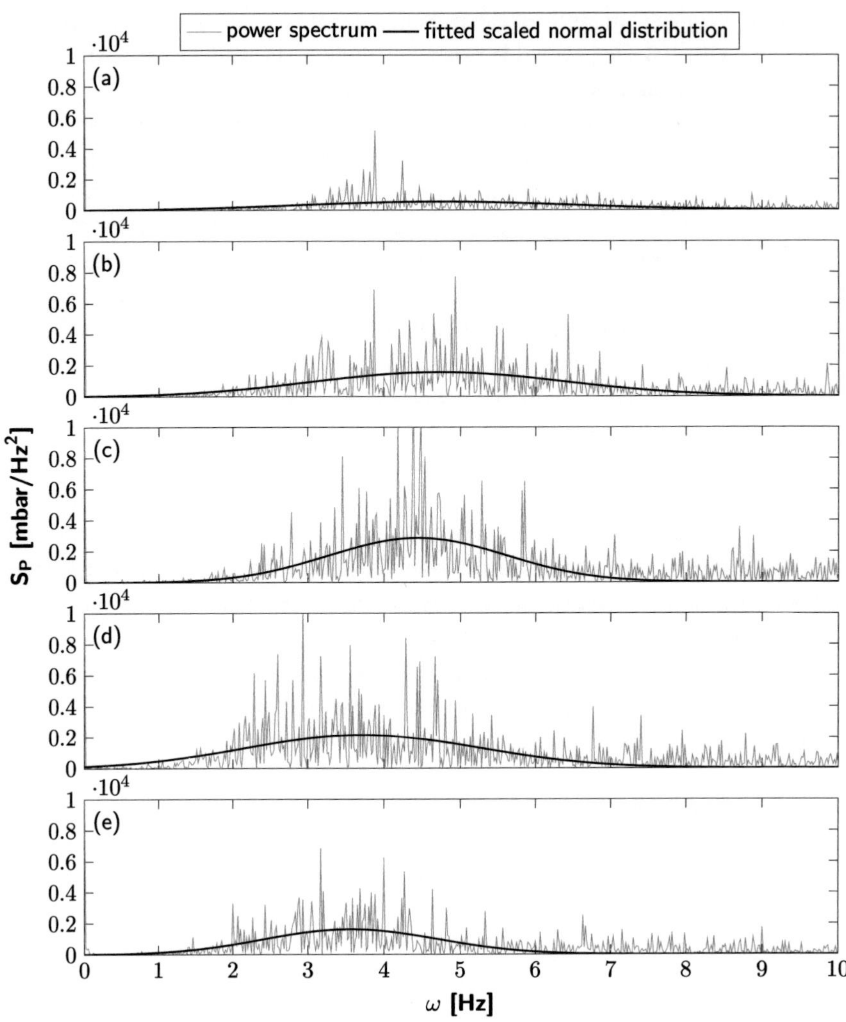

Fig. 4.11: Power spectral density of the pressure signals measured in the windbox with ilmenite at $D = 0.1\,\mathrm{m}$ and $H_0 = 0.15\,\mathrm{m}$ at different superficial gas velocities having values of (a) $U_0 = 0.25\,\mathrm{m\,s^{-1}}$, (b) $U_0 = 0.51\,\mathrm{m\,s^{-1}}$, (c) $U_0 = 0.77\,\mathrm{m\,s^{-1}}$, (d) $U_0 = 1.2\,\mathrm{m\,s^{-1}}$ and (e) $U_0 = 1.58\,\mathrm{m\,s^{-1}}$ (cf. [112]).

For each material, at superficial gas velocities close to minimum fluidization velocity, the mean bed frequency decreases. At larger superficial gas velocities its values vary only slightly in a range of up to 1 Hz without large changes at regime transition. In regard to the determination method of the mean frequency, it can be assumed that above a certain superficial gas velocity the bed reaches a steady state of oscillation. This phenomenon was found for all conditions investigated.

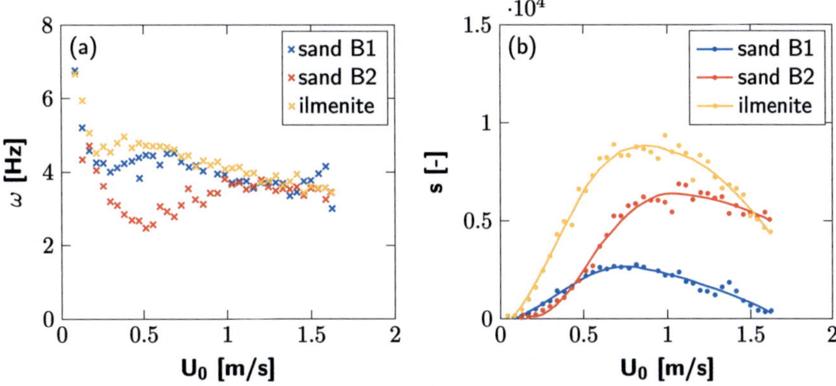

Fig. 4.12: Results of the scaled normal distributions: (a) bed frequencies and (b) scaling parameters of different bed materials ($D = 0.1\,\text{m}$) acc. to [112].

The scaling parameter s of the scaled normal distribution describes the intensity of the power spectral density and gives an approximation of the mean peak heights. The resulting curve for the scaling parameter versus superficial gas velocity (cf. figure 4.12 (b)) shows a similar curvature to the standard deviation of the pressure signal as shown e.g. in figure 4.10 (a). By determination of the maximum of these curves an alternative transition velocity from bubbling to turbulent fluidization called $U_{c,fa}$ is estimated. These transition velocities in comparison to the transition velocities resulting from the evaluation of the standard deviation are shown in a parity plot in figure 4.13. Both determination methods of the transition velocity deliver very similar results with the method by frequency analysis tending to give slightly lower values. For this reason, this method can be used as an alternative way for the determination of the transition velocity from bubbling to turbulent fluidization.

Different studies about the mean frequency of a fluidized bed do exist in literature [202–204]. Baskakov *et al.* [203] describe the fluctuations to be the result of the rising bubbles. Thereby, they compare the oscillation of a fluidized bed with the one of a hydraulic pendulum. Thus, they conclude that the mean frequency of a fluidized bed depends on the bed height at minimum fluidization

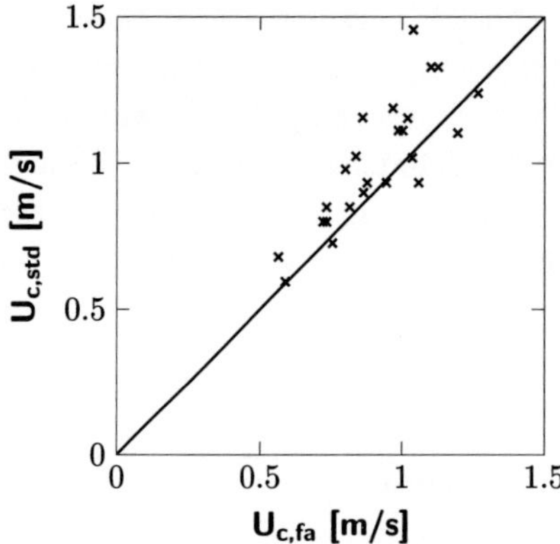

Fig. 4.13: Comparison of the transition velocities from frequency analysis $U_{c,fa}$ and from standard deviation method $U_{c,std}$ (sand B1, sand B2, ilmenite; $D = 0.1\,\text{m}$).

H_{mf} as single operational parameter only. They introduced

$$\omega = \frac{1}{\pi}\sqrt{\frac{g}{H_{mf}}} \tag{4.2}$$

as relationship between the mean bed frequency and the bed height at minimum fluidization.

Results of the determined bed frequencies in comparison to the prediction of equation 4.2 are shown in figure 4.14. In beds having low aspect ratios the bed frequency is found to be larger than in deep beds. A deviation of the measurements from the predicted curve can be seen in the figure. These deviations might result from the determination method. The fitted normal distributions did not always reflect the power spectral density distributions sufficiently leading to an error in the mean frequency of the bed. Furthermore, by comparison of results at diameters of $0.05\,\text{m}$ and $0.1\,\text{m}$ differences in the mean bed frequency can be found at the same static bed heights. Due to the large wall effect fluidization behavior is influenced highly, which might also influence the resulting mean bed frequency.

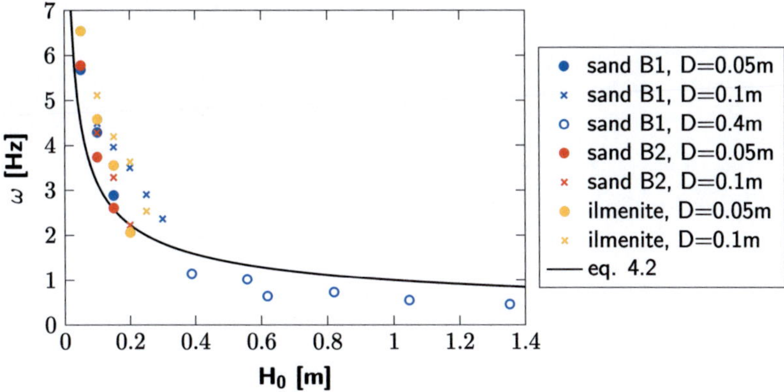

Fig. 4.14: Mean bed frequencies determined in the different facilities in comparison to the prediction by Baskakov *et al.* [203] (assumption: $H_0 = H_{mf}$ for the used materials; cf. [112]).

5 Flow Structure

In this chapter detailed information about the flow structure and the fluid dynamic behavior of a turbulent in comparison to a bubbling fluidized bed is provided. Solids concentrations are measured with capacitance probes and their probabilistic occurrence at variations of parameters like measurement position or superficial gas velocity are analyzed. The information from this is used for the determination of bubble phase hold-ups. Furthermore, bubble properties such as bubble size, velocity, shape and frequency are measured and discussed. Phase hold-ups and bubble properties give information about the gas short-circuiting between bubbles in a fluidized bed. Finally, measurements of the solids entrainment are evaluated and discussed in relation to literature correlations. Some of the results shown in this chapter have been previously published by the author in [123].

5.1 Solids Concentration

In this work, the local solids concentrations in the fluidized bed are measured by capacitance probes. Sections of the calculated solids concentrations from the measurements in the FB100 are shown in figure 5.1 with increasing superficial gas velocity from (a) to (e). The concentrations in the figure are measured in the center of the bed where the main amount of gas bubbles is assumed to rise due to wall effects and preferred bubble flow paths [55]. In figure 5.1 (a) the rise of these bubbles is clearly visible. The concentration of the suspension phase fluctuates at values above a solids concentration of 0.4. Thereby, the solids concentration of the fixed bed is never reached in the figure. This leads to the conclusion that the suspension phase is continuously passed by gas leading to an expansion of this phase. In between the solids concentration sharply drops close to a value of zero. In these cases gas bubbles pass the probe. The concentration does not always drop completely to zero because bubbles can pass the probe only partially or be small in comparison to the measurement volume as discussed in section 3.3.2. Furthermore, the bubble phase can contain small amounts of solids.

With increasing superficial gas velocity in figure 5.1 (b) the visual fluctuations

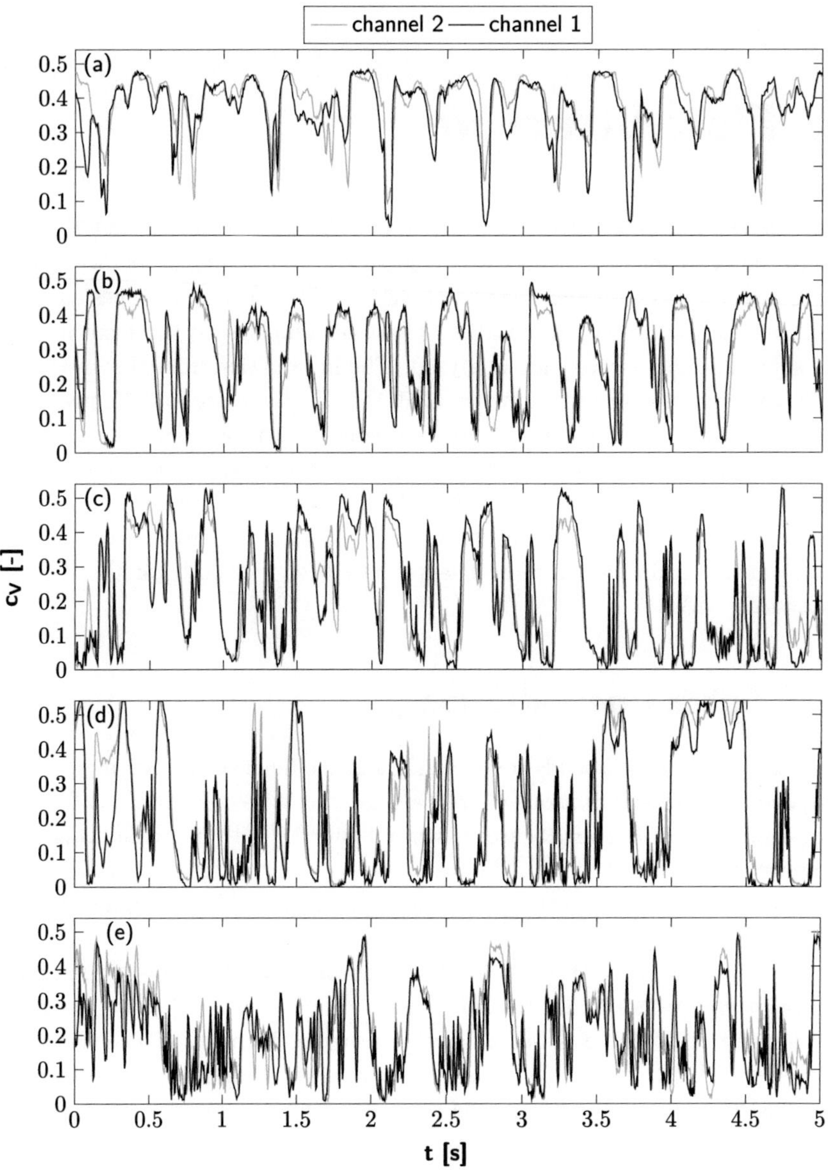

Fig. 5.1: Solids concentration signals of both channels at $H_0 = 0.3\,\text{m}$, $D = 0.1\,\text{m}$, $z = 0.15\,\text{m}$, $r/R = 0$ and (a) $U_0 = 0.18\,\text{m}\,\text{s}^{-1}$, (b) $U_0 = 0.3\,\text{m}\,\text{s}^{-1}$, (c) $U_0 = 0.7\,\text{m}\,\text{s}^{-1}$, (d) $U_0 = 1.1\,\text{m}\,\text{s}^{-1}$ and (e) $U_0 = 1.4\,\text{m}\,\text{s}^{-1}$ (sand B1).

in the concentrations increase, meaning bubbles passing the probe at a higher frequency. This agrees with the expectations that more bubbles having a larger volume rise in the fluidized bed at larger superficial gas velocities.

In figure 5.1 (c) the transition velocity from bubbling to turbulent fluidization is almost reached. The intensity of the fluctuations further increases. Additionally, the solids concentration reaches fixed bed concentration ($c_{V,fb} = 0.54$) in some cases. This might be related to the formation of clusters having large concentrations or to high solids concentrations in bubble wakes or clouds.

In the turbulent regime in figure 5.1 (d) and (e) the solids concentration highly fluctuates. Fixed bed concentration can still be reached but tends to occur less often in the bed center with increasing superficial gas velocity. At large superficial gas velocities electrostatic charging also occurred despite the layering and grounding of the fluidized bed walls with alumina foil. This partially influenced the measurements of the solids concentrations and led to larger fluctuations with sharp peaks.

5.1.1 Local Probability Density Distribution

Local probability density distributions of solids concentrations give information about the absolute occurrence of solids concentrations at a certain measurement point in the fluidized bed. They deliver the distribution of probabilities of concentrations in the bubble phase and the suspension phase if the two phase theory is followed. From this information the bubble phase hold-up can be determined.

A kernel distribution is fitted to the histograms of the probability density distributions. This kernel distribution gives a low error approach for the histograms. Different probability density functions for the diameters of $0.1\,\text{m}$, $0.4\,\text{m}$ and $1\,\text{m}$ in dependence on superficial gas velocity and measurement position are shown in figure 5.2, figure 5.3, and figure 5.4, respectively. For better visualization and due to the low error all graphs contain the fitted distributions only but not the original histograms.

In the probability density distributions the occurrence of low concentrations is generally related to the rise of bubbles through the measurement volume of the probe. Especially at high superficial gas velocities, e.g. in figure 5.2 (e) and (f), large probabilities can be found at very low solid concentrations. The reason for this is the occurrence of large bubbles or large amounts of bubbles. The two phase theory describes bubbles in fluidized beds as gas pockets empty of solid material. In the probability distributions this would lead to peaks exactly at a value of zero. However, coalescence and breakup processes continuously happening can lead to small amounts of solid material inside a bubble. Furthermore, capacitance probes

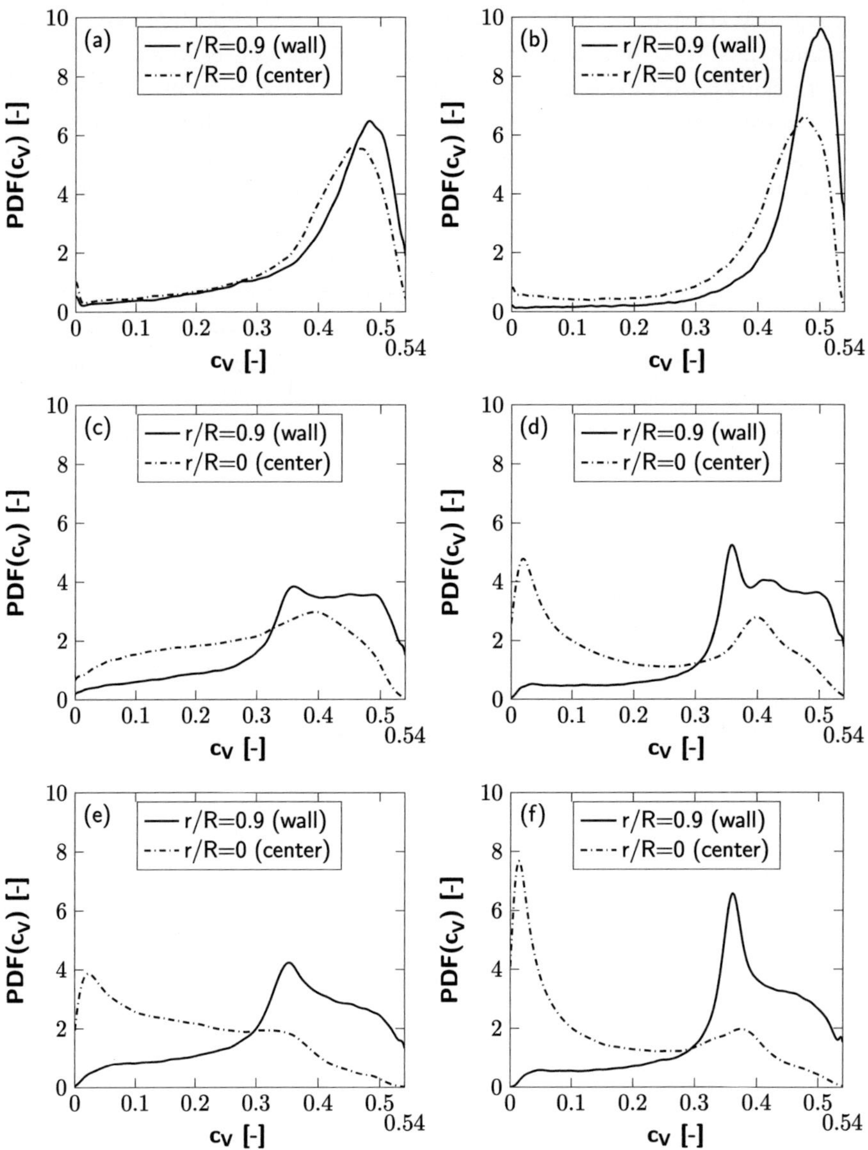

Fig. 5.2: Probability density distributions of the solids concentration of sand B1 for $D = 0.1\,\text{m}$ and $H_0 = 0.2\,\text{m}$ at: (a) $z = 0.05\,\text{m}$, $U_0 = 0.18\,\text{m s}^{-1}$, (b) $z = 0.15\,\text{m}$, $U_0 = 0.18\,\text{m s}^{-1}$, (c) $z = 0.05\,\text{m}$, $U_0 = 0.7\,\text{m s}^{-1}$, (d) $z = 0.15\,\text{m}$, $U_0 = 0.7\,\text{m s}^{-1}$ (cf. [123]), (e) $z = 0.05\,\text{m}$, $U_0 = 1.4\,\text{m s}^{-1}$ and (f) $z = 0.15\,\text{m}$, $U_0 = 1.4\,\text{m s}^{-1}$.

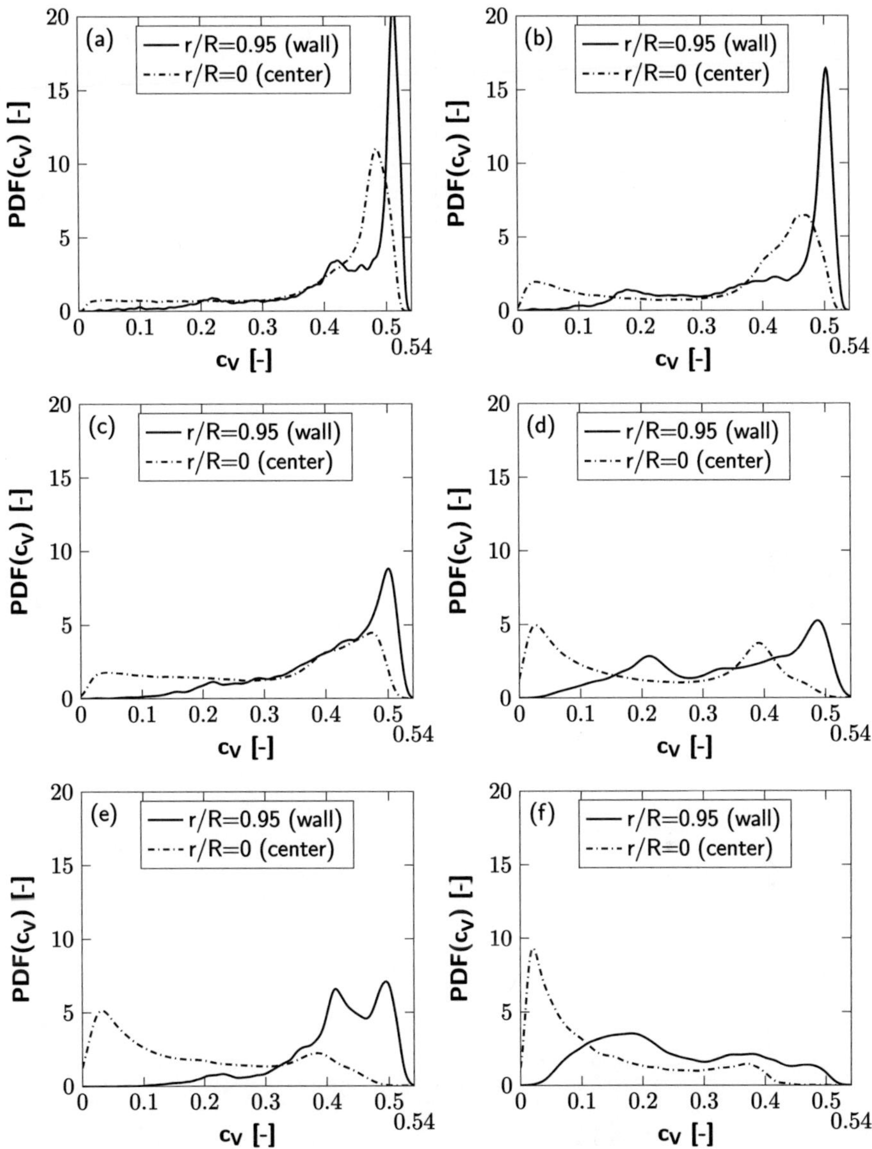

Fig. 5.3: Probability density distributions of the solids concentration of sand B1 for $D = 0.4\,\mathrm{m}$ and $H_0 = 0.4\,\mathrm{m}$ at: (a) $z = 0.19\,\mathrm{m}$, $U_0 = 0.3\,\mathrm{m\,s^{-1}}$, (b) $z = 0.33\,\mathrm{m}$, $U_0 = 0.3\,\mathrm{m\,s^{-1}}$, (c) $z = 0.19\,\mathrm{m}$, $U_0 = 1.25\,\mathrm{m\,s^{-1}}$, (d) $z = 0.33\,\mathrm{m}$, $U_0 = 1.25\,\mathrm{m\,s^{-1}}$, (e) $z = 0.19\,\mathrm{m}$, $U_0 = 2\,\mathrm{m\,s^{-1}}$ and (f) $z = 0.33\,\mathrm{m}$, $U_0 = 2\,\mathrm{m\,s^{-1}}$.

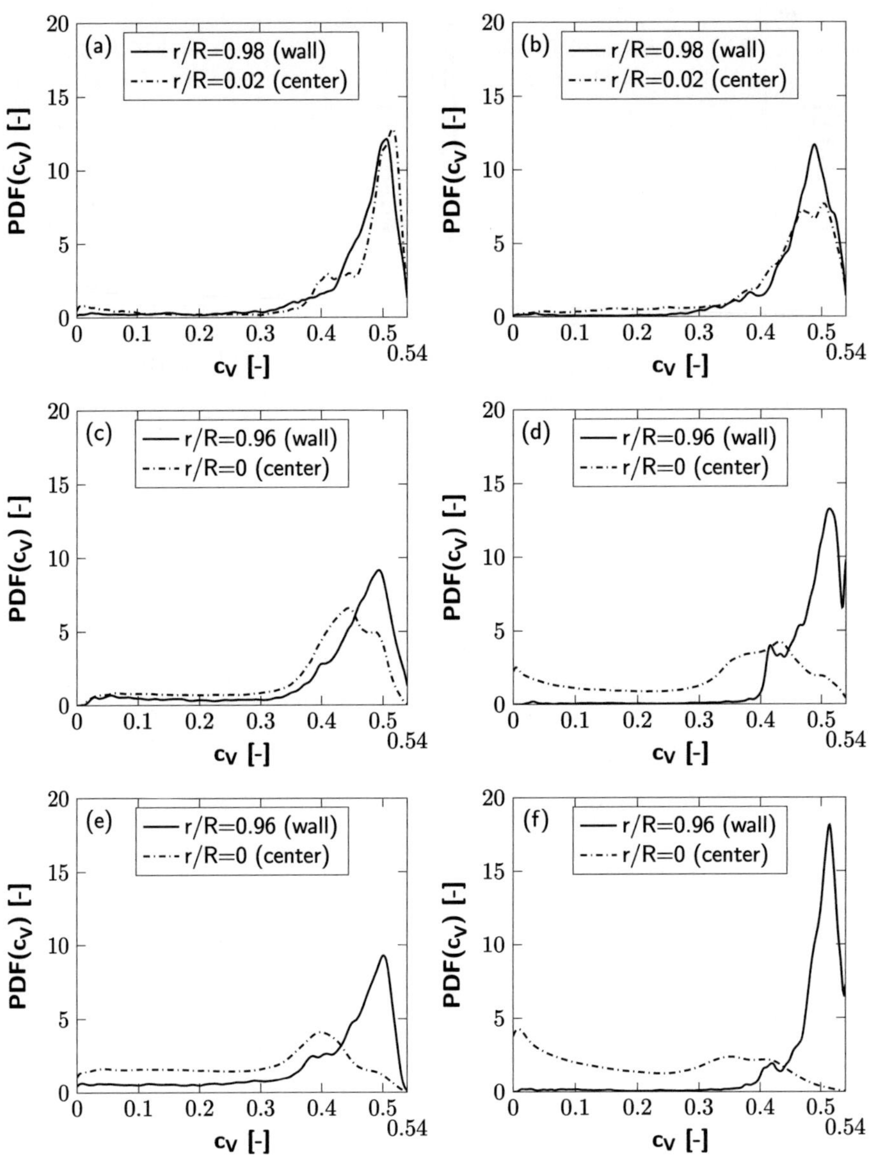

Fig. 5.4: Probability density distributions of the solids concentration of sand B1 for $D = 1\,\text{m}$ and $H_0 = 1\,\text{m}$ at: (a) $z = 0.18\,\text{m}$, $U_0 = 0.3\,\text{m}\,\text{s}^{-1}$, (b) $z = 0.88\,\text{m}$, $U_0 = 0.13\,\text{m}\,\text{s}^{-1}$, (c) $z = 0.18\,\text{m}$, $U_0 = 0.75\,\text{m}\,\text{s}^{-1}$, (d) $z = 0.88\,\text{m}$, $U_0 = 0.75\,\text{m}\,\text{s}^{-1}$, (e) $z = 0.18\,\text{m}$, $U_0 = 1.25\,\text{m}\,\text{s}^{-1}$, and (f) $z = 0.88\,\text{m}$, $U_0 = 1.25\,\text{m}\,\text{s}^{-1}$.

measure in a certain volume defined by the electric field created by the probe. Depending on the probe and bubble size, bubbles can be small in comparison to the measurement volume. In addition, bubbles can pass the probe with the probe penetrating only the phase border of the bubble. In these cases concentrations above a concentration value of zero are measured because both phases exist simultaneously in the measurement volume of the probe.

Peaks occurring at high solid concentrations are the result of the suspension phase passing the probe. Similar to the bubble phase having a solid concentration close to zero, the suspension phase shows a distribution of concentrations close to fixed bed concentration ($c_{V,fb} = 0.54$). The occurrence of a distribution in the concentrations can be explained by the same effects explained above. Furthermore, the solids concentration in the suspension phase must not be related directly to fixed bed concentration because gas streams through the suspension phase leading to partial expansion. In addition to this, falling solids at the wall do not necessarily reach fixed bed concentration.

Figure 5.2 (a) and (b) show one significant peak at high solids concentrations at the wall and in the center, respectively. These peaks represent the suspension phase. Due to the low amount of bubbles rising at this superficial gas velocity of $0.18\,\mathrm{m\,s^{-1}}$ distinct peaks at low concentrations cannot be found. With an increase of the superficial gas velocity the broadness of the distribution of solids concentrations in the suspension phase increases to concentrations below a c_V-value of 0.4 as the graphs of figure 5.2 (c) and (d) show. At these conditions solids concentrations close to fixed bed concentration can only be found close to the fluidized bed wall, whereas in the center of the bed concentrations around a c_V-value of 0.4 show a peak. At a larger distance from the gas distributor a peak close to a c_V-value of zero can be found in the center which represents the rising bubbles. Close to the gas distributor this peak does not occur. This behavior can be explained due to the fact that bubbles rise in preferred flow paths [55]. Bubbles are generated over the whole cross-section of the gas distributor. With increasing height they coalesce towards the center of the bed. From figure 5.2 (c) one can deduce that the main amount of bubbles rises between the wall and the center resulting in the non-existence of peaks at low concentrations at the wall and in the center. In contrast to this, at a larger distance from the gas distributor (5.2 (d)) the main amount of bubbles reaches the fluidized bed center. By further increasing the superficial gas velocity the rate of coalescence is large enough so that bubbles can be observed in the center of the bed at distances to the gas distributor as low as $0.05\,\mathrm{m}$. This can be seen in figure 5.2 (e) in the well developed turbulent fluidized bed regime. The probability of low concentrations in the center of the bed increases further with larger distance from the gas distributor (figure 5.2 (f)). The suspension phase in the turbulent regime shows the occurrence of different concentrations compared to the bubbling regime. Concentrations close to fixed bed concentration can only be found close to the fluidized bed wall. Peaks around a solids concentration of 0.35 can be found at the wall and in the center of the bed as the graphs in figure 5.2 (e) and (f) show. This means a large amount of gas moves also through the suspension phase at a velocity probably larger than minimum fluidization velocity.

Similar behavior can be observed in larger fluidized beds with diameters of 0.4 m (figure 5.3) and 1 m (figure 5.4). In figure 5.3 (a) and (b) low concentrations did not occur close to the fluidized bed wall at both measurement positions above the gas distributor. Bubbles mainly rise in the center of the bed. The probability of the occurrence of low solids concentrations in the center of the bed increases with increasing superficial gas velocity and increasing distance from the gas distributor as the graphs in figure 5.3 show. In contrast to a bed diameter of 0.1 m, at a bed diameter of 0.4 m the distribution of the suspension phase concentrations broadens less intense at larger superficial gas velocities. At the fluidized bed wall solids concentrations close to fixed bed concentration occur with high probabilities also in the well defined turbulent regime. In the center, peaks at a value slightly below a concentration of 0.4 can be found in figure 5.3 (e) and (f) at a superficial gas velocity of $2\,\mathrm{m\,s^{-1}}$. In the graphs of figure 5.4 in the wall region concentrations close to fixed bed concentration can be found at each superficial gas velocity investigated. The same behavior of peak formations in the probability density distributions can be found in this plant as they develop in the others.

5.1.2 Radial and Axial Distribution

To analyze the radial profile of solids concentrations the local measurements are temporally averaged. Results for all three diameters investigated by capacitance probe measurements are shown in figure 5.5 with a diameter of 0.1 m in (a) at bubbling and (b) turbulent fluidization, with a diameter of 0.4 m in (c) at bubbling and (d) turbulent fluidization and with, a diameter of 1 m in (e) at bubbling and (f) also bubbling but at a larger superficial gas velocity.

In the smallest facility (FB100) in the bubbling regime in figure 5.5 (a) the radial solids concentration profile changes significantly with increasing distance from the gas distributor. Close to the distributor a minimum in the concentration can be found between center and wall, whereas at higher distances the lowest concentrations occurred in the bed center. Thus, with distance from the gas distributor the main gas flow moves radially inbound. This behavior is expected due to the description of the formation of a certain flow path behavior in literature as mentioned before [55]. In contrast to this, in the turbulent regime in figure

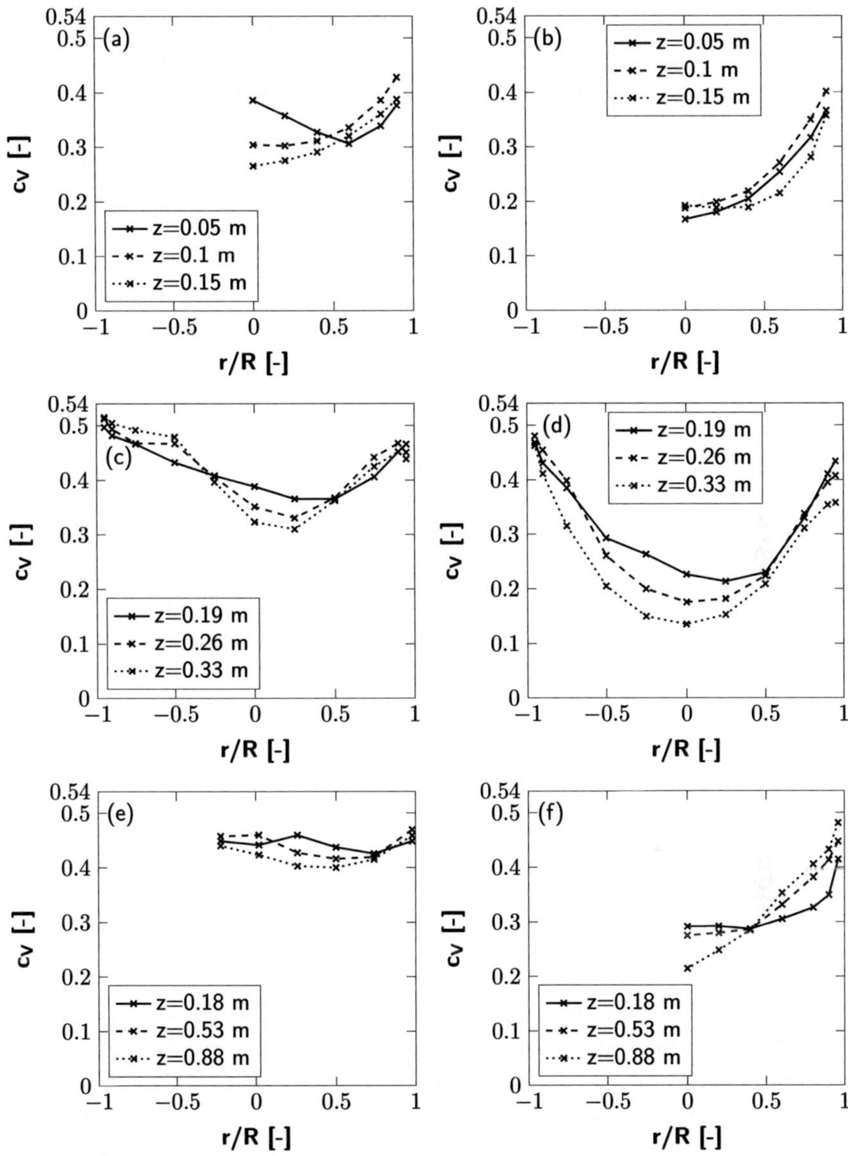

Fig. 5.5: Radial solids concentration profiles for sand B1 at $D = 0.1\,\mathrm{m}$ and $H_0 = 0.3\,\mathrm{m}$ with (a) $U_0 = 0.3\,\mathrm{m\,s^{-1}}$ and (b) $U_0 = 1.4\,\mathrm{m\,s^{-1}}$, at $D = 0.4\,\mathrm{m}$ and $H_0 = 0.4\,\mathrm{m}$ (measurement axis: 180°) with (c) $U_0 = 0.3\,\mathrm{m\,s^{-1}}$ and (d) $U_0 = 1.6\,\mathrm{m\,s^{-1}}$, at $D = 1\,\mathrm{m}$ and $H_0 = 1\,\mathrm{m}$ with (e) $U_0 = 0.3\,\mathrm{m\,s^{-1}}$, and (f) $U_0 = 1.25\,\mathrm{m\,s^{-1}}$.

5.5 (b) the solids concentrations profiles seem to be very similar and almost independent of height in the dense zone of the fluidized bed.

In the CFB400 in the bubbling regime (see figure 5.5 (c)) the development of preferred gas flow from the wall to the center could not be observed. This might be related to the positioning of the lowest measurement being at the point of the main gas flow reaching the bed center. At larger distances from the gas distributor in the range of -1 to -0.5 of the dimensionless radius almost constant concentrations close to fixed bed concentration can be found. The profiles do not show axial symmetry and the minimum of the concentration profile is slightly shifted away from the bed center. This behavior diminishes in the turbulent regime in figure 5.5 (d). The profiles formed are of parabolic shape and the minimum mean solids concentration decreases with increasing distance from the gas distributor. This effect occurs due to the static bed height being at a value of $0.4\,\mathrm{m}$ and the expansion of the bed leading to the upper measurements being in the transition zone of the lower dense zone and the upper freeboard zone of the bed.

In the largest facility similar behavior as in the FB100 can be found in the bubbling state in figure 5.5 (e). At each height investigated the concentration minimum was not found to be in the center of the bed as it is the case close to the gas distributor in the FB100. Thus, it can be concluded that the formation of two circulation cells occurred as shown schematically in figure 2.4. Two main flows of rising gas occur and solids descend at the fluidized bed wall and in the center of the bed. This flow behavior changes if the superficial gas velocity is increased. Flow profiles similar to the parabolic one occur if the superficial gas velocity is set close to the transition into the turbulent regime as shown in figure 5.5 (f).

In figure 5.6 measurements of the solids concentration at a constant distance from the gas distributor but from different angles are plotted in (a) and (b) at two different superficial gas velocities. The measurements slightly differ from each other. Complete axial symmetry is not given, which can be related to the location of the solids return and the nature of the fluidized bed to form regions of preferred bubble flow. Furthermore, an uneven gas distribution can affect the flow of bubbles.

To compare the solids concentration measurements with pressure measurements the temporally averaged concentrations are averaged over the bed cross-section. Using equation 2.5 the solids concentrations predicted from the pressure drop measurements are calculated. The differential pressures are measured between two different heights and yield a solids concentration averaged over time, the bed cross-section, and the investigated height segment. Axial profiles of both different kinds of solids concentrations, determined by probe and by pressure,

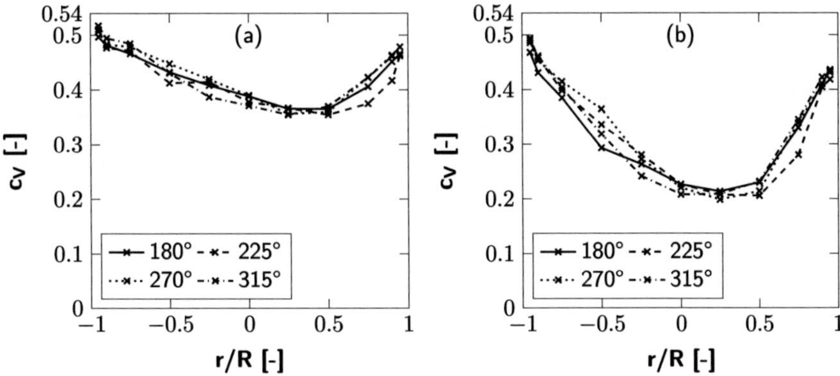

Fig. 5.6: Radial solids concentration profiles at (a) $U_0 = 0.3\,\mathrm{m\,s}^{-1}$ and (b) $U_0 = 1.6\,\mathrm{m\,s}^{-1}$ measured at different angles ($D = 0.4\,\mathrm{m}$, sand B1).

are compared in figure 5.7 for a bed diameter of 0.4 m at two different static bed heights and at varying superficial gas velocities. At a low static bed height in figure 5.7 (a) the solids concentration calculated from the pressure measurement is larger than the one determined by capacitance probe up to a height of about 0.3 m. In addition to this, the solids concentrations calculated from pressure highly fluctuate up to this height in contrast to the solids concentrations measured with the capacitance probe. The same behavior can be observed at a larger static bed height in figure 5.7 (b). There, in the dense zone of the bed, the solids concentration measured with the capacitance probe is constant, whereas the one obtained by pressure measurements fluctuates between c_V-values of 0.4 and fixed bed concentration ($c_{V,fb} = 0.54$). An overestimation of the solids concentration from pressure measurements can be observed generally in the dense zone of the fluidized bed. Such behavior results from acceleration phenomena in this zone of the fluidized bed leading to a larger pressure drop, also reported in literature [110, 205]. Furthermore, pressure measurements are recorded at the fluidized bed wall in this work and local fluid dynamic influences on the pressure measurement cannot be excluded.

Above a certain height in the bed, where the transition between dense zone and freeboard occurs and the solids concentration drops, the solids concentration determined by pressure significantly drops below the solids concentration by capacitance probe. This behavior can be observed for larger superficial gas velocities in figure 5.7 (a). It can be explained by deceleration of solids leading to the opposite effect that occurs in the dense zone. The pressure drop measured in this zone is decreased from the decelerated solids.

Because in the FB1000 it was not possible to reach the turbulent flow regime,

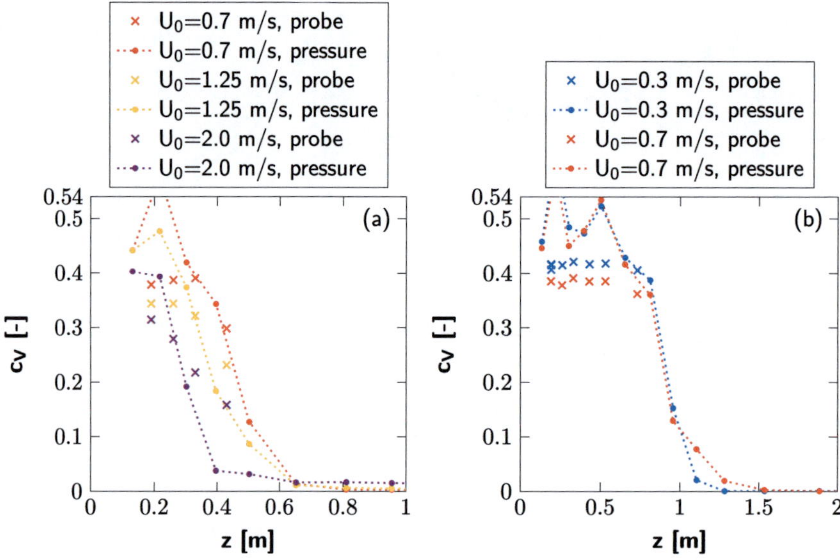

Fig. 5.7: Axial solids concentration profiles measured for sand B1 by capacitance probe in comparison to the determination by pressure measurement at $D = 0.4\,\text{m}$ and (a) $H_0 = 0.4\,\text{m}$ and (b) $H_0 = 0.8\,\text{m}$.

large influences by acceleration and deceleration of solids cannot be observed. Nevertheless, in the dense zone of the bed solids concentrations by pressure measurements slightly overestimate the actual solids concentration, which is determined by capacitance probe. This behavior can be observed in figure 5.8.

5.1.3 Phase Definition

To determine the phase holdups, the bubble and suspension phase must be defined. According to literature phase definition can be done by evaluation of the probability density distributions of the solids concentrations [55]. Using the `findpeaks` function in MATLAB®, peaks of bubble and suspension phase in the probability density distributions are determined in this work. In the suspension phase multiple peaks occurred simultaneously especially in the turbulent regime (see e.g. figure 5.3 (e)). These peaks are manually allocated to two different subphases in the suspension phase called low and high concentration suspension phases. Peaks of the high concentration suspension phase close to fixed bed concentration mainly occurred in the bubbling regime and in the wall region of the fluidized bed as discussed above. In contrast to this, peaks categorized to the low concentration suspension phase have concentration values in the range

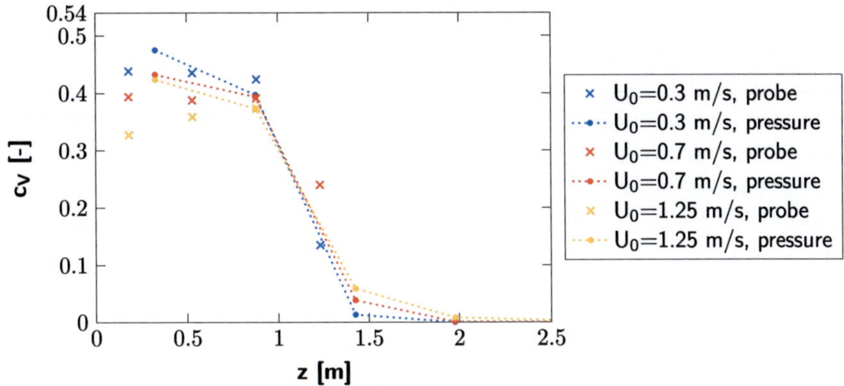

Fig. 5.8: Axial solids concentration profiles measured for sand B1 by capacitance probe in comparison to the determination by pressure measurement at $D = 1\,\text{m}$ and $H_0 = 1\,\text{m}$.

of 0.4 to 0.45 and are found more likely in the center and at larger superficial gas velocities. Peaks found in the bubble phase are always related to the rise of bubbles in the fluidized bed. Evaluation showed that location of the peaks is independent on radial and axial measurement position. For this reason, the peak concentrations are averaged cross-sectionally and axially.

Results for the averaged peak concentrations for suspension phase and bubble phase at bed diameters of $0.4\,\text{m}$ and $1\,\text{m}$ are shown in figure 5.9. Both, peak concentrations of the low and the high concentration suspension phase show a dependence on the superficial gas velocity. With increasing superficial gas velocity the suspension phase concentrations decrease linearly. This is the result of larger amounts of gas streaming through the suspension phase. As a linear approach for the determination of the solids concentrations of the high concentration suspension $c_{V,S,high}$ and the low concentration suspension of sand B1 equations 5.1 and 5.2 are fitted to the data in dependence on the superficial gas velocity U_0:

$$c_{V,S,high} = -0.018 U_0 + 0.51, \tag{5.1}$$

$$c_{V,S,low} = -0.033 U_0 + 0.454 \tag{5.2}$$

Concentrations of the peaks determined in the bubble phase are found to be independent of superficial gas velocity as given by

$$c_{V,B} = 0.029. \tag{5.3}$$

Results in the fluidized bed having a diameter of $0.1\,\text{m}$ are not shown in figure

5.9 because a significant influence of the bed diameter on the suspension phase concentrations was found in this case. In the low concentration suspension phase as well as in the high concentration suspension phase lower concentrations are measured for the smallest bed investigated than in the larger ones. This happens due to the high interaction of solids and gas at small diameters resulting in more gas streaming through the dense phase. In contrast, in a large fluidized bed preferred flow paths of gas and solid are developed leading to lower interaction of the phases.

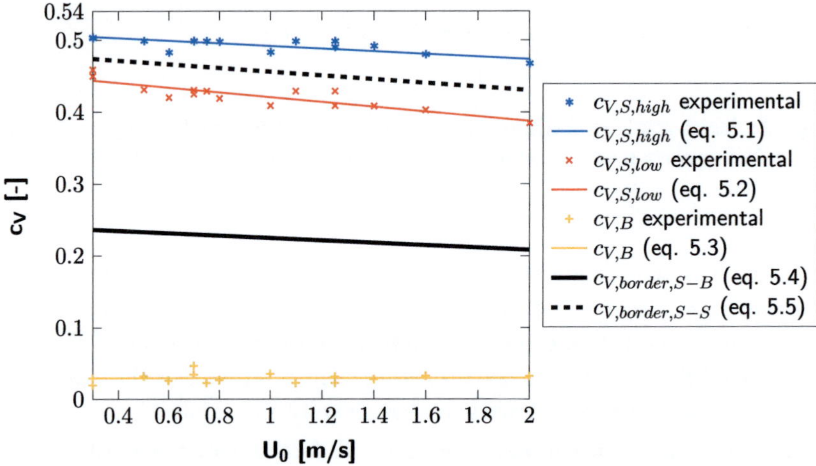

Fig. 5.9: Cross-sectional and axial averaged peak concentrations of suspension and bubble phase and the resulting phase definition equations 5.4 and 5.5 for sand B1.

To distinguish between bubble and dense phase the phase border concentration is defined as mean value between low concentration suspension phase and bubble phase concentration resulting in

$$c_{V,border,S-B} = -0.017U_0 + 0.242 \tag{5.4}$$

for sand B1 as it is shown in figure 5.9. The resulting concentrations agree well with the minimums that can be found in the probability density distributions between both phases. This border definition is used to determine the phase holdups and bubble properties. In addition, the subphases of the suspension phase for sand B1 are divided by

$$c_{V,border,S-S} = -0.026U_0 + 0.482 \tag{5.5}$$

resulting from the mean value of low and high concentration suspension holdup.

5.2 Phase Holdup

As defined in section 5.1.3 bubble and suspension phase are distinguished by equation 5.4. The probability density distributions give information about probability of the occurrence of local concentrations. Thus, by the deviation of the integral of the probability density distribution from zero to phase border concentration $c_{V,border,S-B}$ to the integral of the whole probability density function from zero to fixed bed concentration $c_{V,fb}$ the bubble phase holdup can be calculated according to

$$\phi_B = \frac{\int_0^{c_{V,border,S-B}} PDF\,(c_V)\,dc_V}{\int_0^{c_{V,fb}} PDF\,(c_V)\,dc_V}.$$ (5.6)

Analog to this the low concentration suspension phase fraction $\psi_{S,low}$ is calculated by:

$$\psi_{S,low} = \frac{\int_{c_{V,border,S-B}}^{c_{V,border,S-S}} PDF\,(c_V)\,dc_V}{\int_{c_{V,border,S-B}}^{c_{V,fb}} PDF\,(c_V)\,dc_V}$$ (5.7)

Both parameters are related by the phase balance:

$$1 = \phi_B + \phi_S = \phi_B + \left(\psi_{S,low} + \psi_{S,high}\right)\phi_S$$ (5.8)

Thereby, ϕ_S defines the suspension phase holdup and $\psi_{S,high}$ the high concentration suspension phase fraction.

Radial and axial phase holdup profiles can be received and evaluated. Latter can be obtained by cross-sectional averaging of the radial profiles at different axial positions.

5.2.1 Radial Phase Distribution

The radial phase distribution of the bubble phase holdup gives information about preferred flow paths of bubbles in a fluidized bed. In figure 5.10 the bubble phase holdups under different conditions (different distances from gas distributor and superficial gas velocities) in the three plants investigated are shown. The radial profiles in the plants having a diameter of 0.1 m and 1 m have been measured over the half bed diameter, whereas at a diameter of 0.4 m the whole diameter was investigated, as mentioned in chapter 3.

The bubble phase holdup profiles in the smallest plant (FB100) at bubbling (figure 5.10 (a)) and turbulent (figure 5.10 (b)) fluidization show the distinctive flow behavior of bubbles described in sections 2.2 and 2.3. At a low superficial gas velocity bubbles coalesce with increasing distance from the gas distributor from the fluidized bed wall into the center of the fluidized bed. This results in a peak at a dimensionless radius of 0.6 at a distance of 0.05 m above the gas distributor, whereas a minimum occurs directly at the wall and in the bed center (assuming axial symmetry). Most of the bubble volume rises at the peak point and solid recirculates at the wall and in the bed center [55]. At heights of 0.1 m and 0.15 m the main amount of bubbles already reached the center of the bed resulting in a peak in the fluidized bed center. In turbulent fluidization a parabolic profile is developed also at distances close to the gas distributor. Gas in the bubble phase rises mainly in the center of the bed. The backflow of solids mainly occurs at the fluidized bed wall resulting in no bubbles rising in this region of the bed. This is shown by the curves approaching a value of zero at a dimensionless radius of one in figure 5.10 (b).

At a larger bed diameter of 0.4 m in the well developed turbulent regime (figure 5.10 (d)) the formation of the parabolic profile with no bubbles rising at the wall can be observed as well. In this case the bubble phase holdup in the bed center increases with distance from the gas distributor. This can be related to a static bed height set to a value of 0.4 m for these experiments. In the turbulent regime particles can be entrained due to their low terminal velocities compared to the gas velocities occurring in the bed. A smooth transition between the lower dense phase and the upper dilute phase of the bed is formed. The measurements at heights close to the static bed height (like $z = 0.33$ m) are influenced by this transition already resulting in lower amounts of solids and higher bubble phase holdups. In figure 5.10 (c) in the bubbling regime the movements of bubbles into the bed center with increasing height can be found similar to 5.10 (a). At a height of 0.19 m most bubbles reached the bed center. With increasing distance from the gas distributor the bubbles coalesce further into the bed center resulting in a bubble phase holdup profile with very few bubbles in the regions close to the wall and high values in the center as shown in the figure. Due to the construction of the plant, the backflow of solids via the loop-seal and the nature of the fluidized bed the profiles are not completely axially symmetric.

At a diameter of 1 m in the bubbling regime at a superficial gas velocity of $0.3 \, \mathrm{m\,s^{-1}}$ peaks of the bubble phase holdup in the range of the dimensionless diameter of 0.25 to 0.5 can be found at heights of 0.53 m and 0.88 m. Under the assumption of axial symmetry, two preferred flow paths of bubble rise are formed as it was found in literature for large plants as well [55]. Interaction of bubbles is limited due to the size of the plant and solid material flows back to where the occurrence of bubbles is improbable. With increasing gas velocity the total bubble holdup increases leading to the merge of bubbles in the center of the bed as shown by figure 5.10 (f).

At an aspect ratio (static bed height divided by bed diameter) of 2 in the

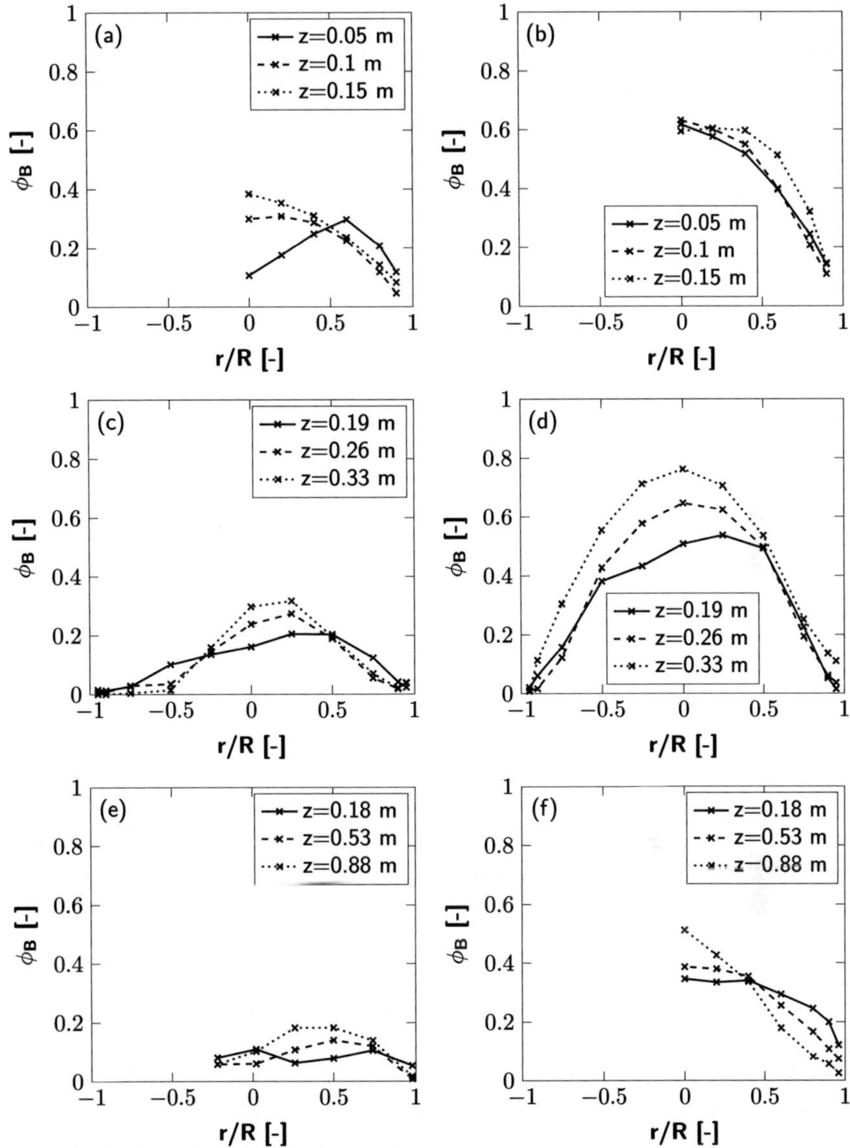

Fig. 5.10: Radial suspension phase hold-up profiles for sand B1 at $D = 0.1\,\mathrm{m}$ and $H_0 = 0.3\,\mathrm{m}$ with (a) $U_0 = 0.3\,\mathrm{m\,s^{-1}}$ and (b) $U_0 = 1.4\,\mathrm{m\,s^{-1}}$, at $D = 0.4\,\mathrm{m}$ and $H_0 = 0.4\,\mathrm{m}$ (measurement axis: 180°) with (c) $U_0 = 0.3\,\mathrm{m\,s^{-1}}$ and (d) $U_0 = 1.6\,\mathrm{m\,s^{-1}}$, at $D = 1\,\mathrm{m}$ and $H_0 = 1\,\mathrm{m}$ with (e) $U_0 = 0.3\,\mathrm{m\,s^{-1}}$, and (f) $U_0 = 1.25\,\mathrm{m\,s^{-1}}$.

0.4 m diameter plant the bubble phase holdup in the center was found to increase slightly with increasing height as can be seen in figure 5.11. By cross-sectional averaging the bubble phase holdups, it was found that the mean bubble phase holdup is constant and thus independent of the height above the gas distributor. For this reason, the phase holdup behavior observed in figure 5.11 is the result of bubbles coalescing into the fluidized bed center. At the same time, the occurrence of bubbles in the wall region gets less probable with increasing distance from the gas distributor.

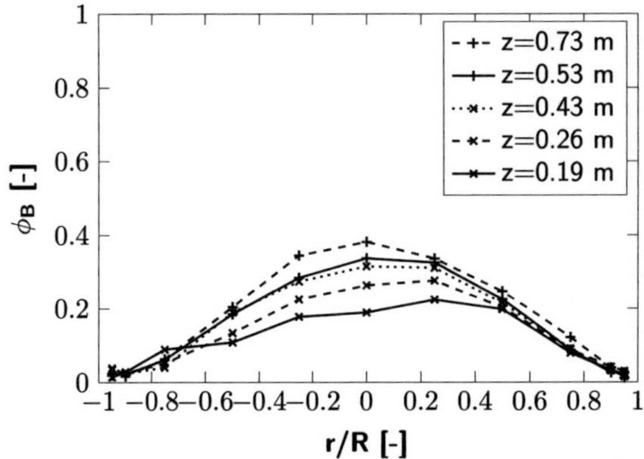

Fig. 5.11: Suspension phase hold-up dependence of height above the gas distributor for sand B1 at $D = 0.4$ m, $H_0 = 0.8$ m and $U_0 = 0.3$ m s^{-1} (bubbling regime).

In figure 5.12 the dependence of the bubble phase holdup on the superficial gas velocity is shown for the 0.4 m diameter bed at heights above the gas distributor of (a) 0.19 m and (b) 0.33 m. As discussed before, the measurement height of 0.33 m is close to the static bed height of 0.4 m that was set for these experiments and at large superficial gas velocities the results show phase holdups in the transition zone between the dense and dilute zone of the fluidized bed. This leads to large bubble phase holdups for superficial gas velocities of 1.6 m s^{-1} and 2 m s^{-1} in figure 5.12 (b). In the dense zone of the fluidized bed (figure 5.12 (a)) the bubble phase holdup approaches a value of zero at the wall of the fluidized bed independent of superficial gas velocity. In the center of the bed the bubble phase holdup increases continuously at higher superficial gas velocities meaning more gas is streaming through the bed in the bubble phase. In bubbling fluidization zones of deprived bubble flow can be observed as discussed before and shown in figure 5.12 (b) at a superficial gas velocity of 0.3 m s^{-1} with almost no bubbles rising in the wall region. In contrast to this, turbulent fluidization is defined by parabolic bubble holdup profiles. Rising gas influences the suspension phase at each point in the bed and the occurrence of fluid dynamic dead zones is improbable. Among other things, this leads to good mixing and the favorable high heat and mass transfer in turbulent fluidized beds.

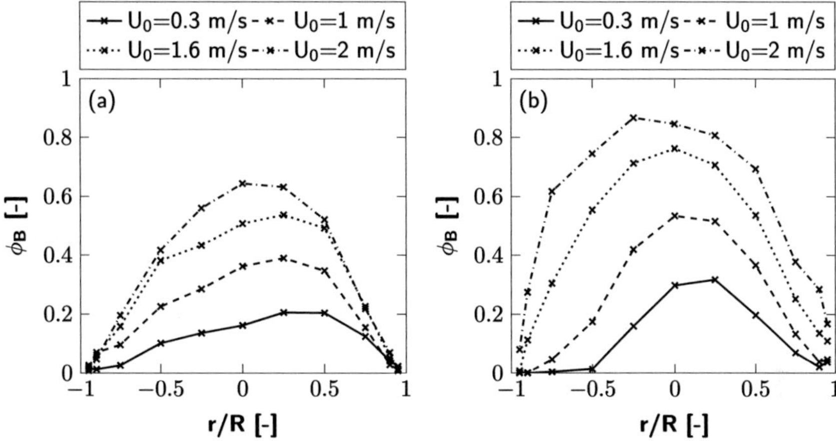

Fig. 5.12: Suspension phase hold-up dependence of superficial gas velocity for sand B1 at $D = 0.4\,\mathrm{m}$ and $H_0 = 0.4\,\mathrm{m}$ at heights above the gas distributor of (a) $z = 0.19\,\mathrm{m}$ and (b) $z = 0.33\,\mathrm{m}$.

The fluidized bed diameter has a large impact on bubble phase holdup in small plants. In figure 5.13 the bubble phase holdups of the three plants investigated (a) in bubbling fluidization and (b) at the transition to turbulent fluidization under similar conditions are shown. Independent of superficial gas velocity, in the smallest plant of the three plants the bubble phase holdup is the largest in the bed center. Bubbles in small beds can reach a size in the order of magnitude of the fluidized bed diameter fast. Thus, flow interaction between bubble and suspension phase is large and the rising bubbles can hinder solids from falling down at the fluidized bed wall. The result is a larger expansion of the bed and higher bubble phase holdups. This influence decreases with increasing size of the bed. In a large bed the formation of preferred flow paths and bubbles rising in chains [55, 79] reduce phase flow interactions because regions of bubble rise and solid backflow are formed. Furthermore, in larger beds bubbles can attain larger sizes. This results in larger rise velocities and a lower residence time in the bed. For this reason, phase holdups in the fluidized bed plant having a diameter of 1 m are much lower than at 0.1 m as it is shown in figure 5.13.

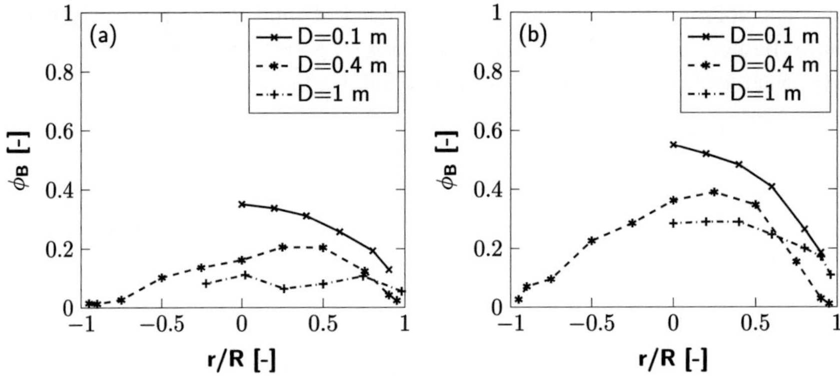

Fig. 5.13: Suspension phase hold-up at different plant diameters ($D = 0.1\,\mathrm{m}$ measured at $H_0 = 0.3\,\mathrm{m}$, $z = 0.2\,\mathrm{m}$; $D = 0.4\,\mathrm{m}$ measured at $H_0 = 0.4\,\mathrm{m}$, $z = 0.19\,\mathrm{m}$, $D = 1\,\mathrm{m}$ measured at $H_0 = 1\,\mathrm{m}$, $z = 0.18\,\mathrm{m}$) and superficial gas velocities for sand B1 at (a) $U_0 = 0.3\,\mathrm{m\,s^{-1}}$ and (b) $U_0 = 1.1\,\mathrm{m\,s^{-1}}$ (for $D = 0.1\,\mathrm{m}$ and $D = 1\,\mathrm{m}$), and $U_0 = 1\,\mathrm{m\,s^{-1}}$ (for $D = 0.4\,\mathrm{m}$).

5.2.2 Cross-sectionally averaged Phase Distribution

Detailed two dimensional modeling of the bubble phase holdup would need much more data at different diameters, static bed heights and preferably also different materials than gathered in this work. For this reason, radial influences are neglected by cross-sectional averaging of the bubble phase holdup to create a one dimensional empiric correlation that predict the holdups. Axial profiles of the bubble phase holdups at bed diameters of $0.4\,\mathrm{m}$ and $1\,\mathrm{m}$ at different superficial gas velocities are given in figure 5.14. As discussed before, the average bubble phase holdup in the dense phase of the fluidized bed is found to be independent of the distance from the gas distributor. With increasing superficial gas velocity the transition into the upper dilute zone of the bed gets smoother and bubble phase holdups increase as a result of increasing dilution as shown in figure 5.14 (a). The cross-sectional averaged bubble phase holdup in the dense phase is found to increase continuously at higher superficial gas velocities. This behavior is expected because at higher superficial gas velocities generally larger amounts of gas need to pass through the bed and due to the resistance and inertia of the solid material this does not only result in larger bubble velocities.

Equation 5.9 was fitted to the measurement data of the bubble phase holdups with sand B1 in dependence on superficial gas velocity and bed diameter under the assumption of independence from the static bed height (all parameters in SI

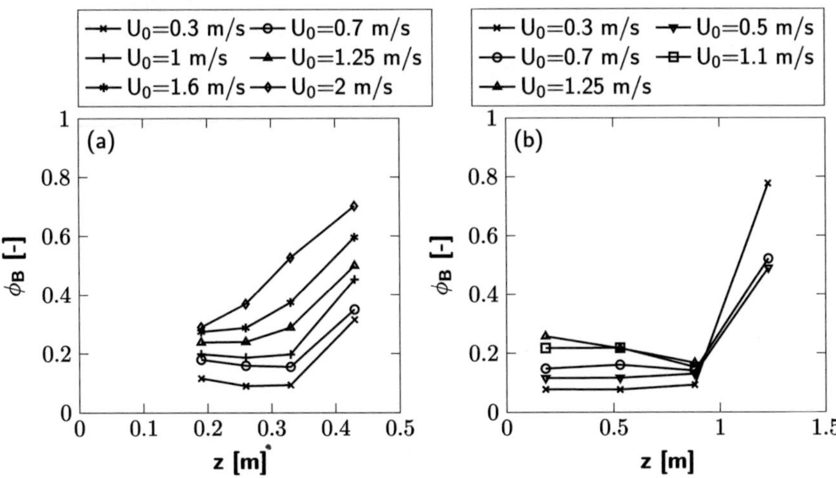

Fig. 5.14: Cross-sectionally averaged suspension phase hold-ups in dependence on height above the gas distributor and superficial gas velocity for sand B1 at different bed diameters (a) $D = 0.4\,\mathrm{m}$, $H_0 = 0.4\,\mathrm{m}$ and (b) $D = 1\,\mathrm{m}$, $H_0 = 1\,\mathrm{m}$.

units; cf. [123]):

$$\phi_B = \frac{1}{1 + \left(\frac{4.735D}{0.128+D}\right)(U_0 - U_{mf})^{-0.641}} \tag{5.9}$$

Due to inconsistent data regarding the influence of the distance from the gas distributor on the bubble phase holdup an direct influence was not considered in the equation above.

For large beds the influence of bed diameter becomes negligible resulting in

$$\phi_{B,D\to\infty} = \frac{1}{1 + 4.735\,(U_0 - U_{mf})^{-0.641}}, \tag{5.10}$$

with all parameters in SI units. The diameter independence for large beds is a result of the size of bubbles being small in comparison to the fluidized bed diameter. Thus, the bubbles and their rise are less influenced by the fluidized bed wall and wall effects decrease.

Results of the bubble phase holdup in comparison to the fitted curves are shown in figure 5.15. The predictions of the correlation are in a agreement with the measurement with most of them being in a band of $20\,\%$ as shown in the parity graph in figure 5.16. The largest errors are found to occur mainly at the smallest bed diameter. This is related to the size of bubbles that can reach the order of magnitude of the size of the bed fast, especially at elevated heights above the gas distributor in these beds. Thus, flow behavior is influenced due to this as discussed above. Furthermore, influences of the distance from the gas distributor might play a role especially at small fluidized bed diameters.

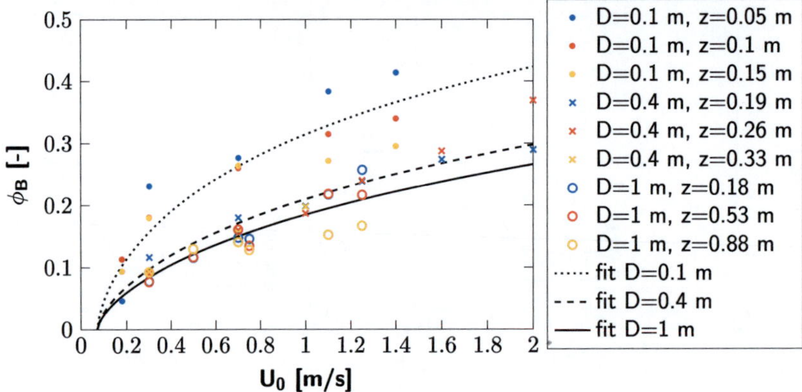

Fig. 5.15: Cross-sectionally averaged bubble phase hold-ups for sand B1 in dependence on superficial gas velocity at different bed diameters ($D = 0.1\,\mathrm{m}$, $D = 0.4\,\mathrm{m}$, $D = 1\,\mathrm{m}$) and different heights above the gas distributor with their predicted curves fitted from equation 5.9 acc. to [123].

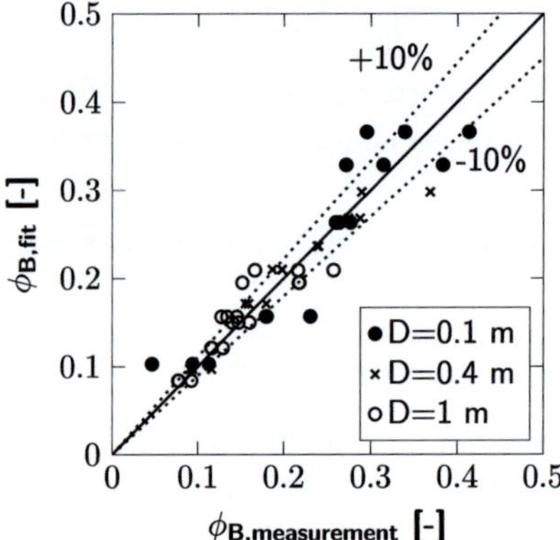

Fig. 5.16: Parity plot of calculated and measured bubble phase holdups; calculation by equation 5.9 acc. to [123].

In addition to the definition of bubble and suspension phase, the suspension phase is subdivided into the low and the high concentration suspension phase as described above. It is found that high concentrations close to fixed bed concentration mainly occur at the fluidized bed wall, whereas the low concentration suspension phase is dominant in the center of the fluidized bed. The results for the cross-sectionally averaged low concentration suspension phase fraction (defined in equation 5.7) are shown in figure 5.17. A clear dependence of the distance from the gas distributor is not found as it is the case for the bubble phase holdup. Furthermore, the influence of superficial gas velocity is low for larger velocities. With increasing superficial gas velocity the occurrence of lower concentrations in the suspension phase gets more probable. The low concentration suspension phase might be related to wakes and clouds of bubbles where concentrations are known to be lower than at minimum fluidization velocity. This explains the similarity in the radial behavior to the bubble phase holdup. Furthermore, at larger superficial gas velocities more bubbles with larger wakes occur resulting in the higher probability of the low concentration suspension phase.

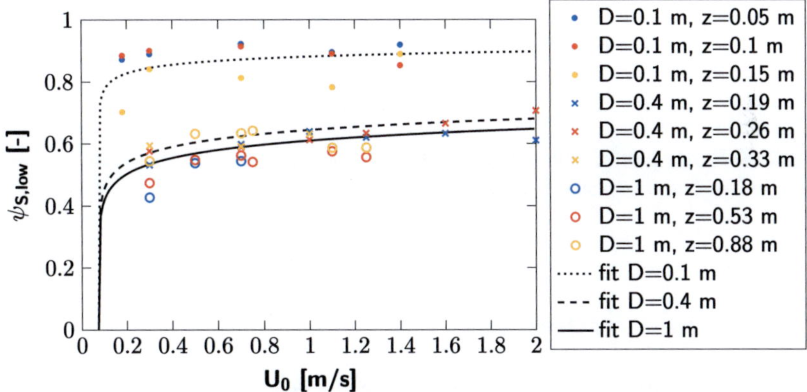

Fig. 5.17: Cross-sectionally averaged holdups of low concentration suspension phase for sand B1 in dependence of superficial gas velocity at different bed diameters ($D = 0.1\,\mathrm{m}$, $D = 0.4\,\mathrm{m}$, $D = 1\,\mathrm{m}$) and different heights above the gas distributor correlated.

The influence of bed diameter is significant as shown in figure 5.17. The occurrence of concentrations close to fixed bed concentration gets less probable at lower diameters due to phase interactions. The measurement data of sand B1 is fitted by a correlation (all parameters in SI units):

$$\psi_{S,low} = \frac{1}{1 + \left(\frac{12.183D}{0.0048+D} - 11.497\right)(U_0 - U_{mf})^{-0.22}} \tag{5.11}$$

The influence of the bed diameter is limited resulting in

$$\psi_{S,low,D\to\infty} = \frac{1}{1 + 0.686\,(U_0 - U_{mf})^{-0.22}} \tag{5.12}$$

for large plants (all parameters in SI units). The parity graph in figure 5.18 shows errors up to 20 %. These errors are related to the definition of the low and high concentration suspension phase fraction and border, which can deviate due to different fluid dynamic behavior in bed of different sizes.

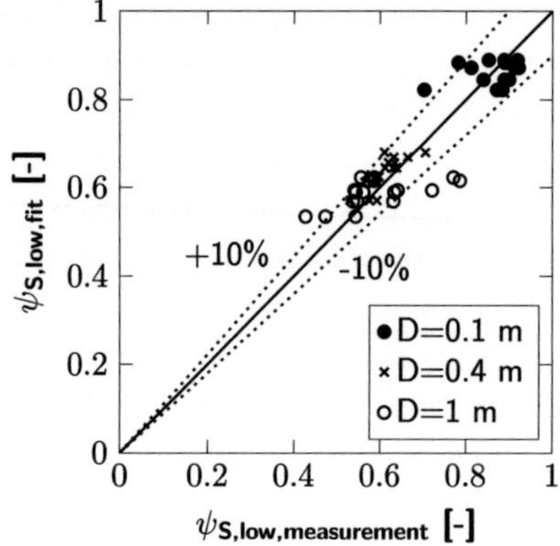

Fig. 5.18: Parity plot of calculated hold-ups of low concentration suspension phase by equation 5.11 and measured data.

5.3 Bubble Properties

For a deeper understanding of the change of the flow behavior between bubbling and turbulent fluidization bubble properties such as bubble size and velocity have been determined. These parameters are crucial for heat and mass transfer in fluidized beds.

5.3.1 Bubble Size

The movement of bubbles in fluidized beds is much larger in vertical direction than laterally. Measurement techniques relying on the principle of measurement of temporal local changes of the solids concentration predominantly allow the measurement of vertical bubble properties. A change in concentration is more likely to happen due to a bubble rising into or out of the measurement volume of the probe than by a lateral movement. Nevertheless, lateral movements occur due to coalescence phenomena as reported in literature [55]. The capacitance probes used in this work are designed to determine the vertical sizes of bubbles (pierced length) and bubble velocities in vertical direction.

As described in section 3.3.2 the pierced length measured is always a result of the overlapping distributions of bubble size and location where the probe penetrates the bubble. Probability density distributions of the pierced lengths measured in the plant having a diameter of 0.1 m at three different superficial gas velocities and two different measurement heights above the gas distributor are shown in figure 5.19. The histograms are fitted with a logarithmic normal distribution, which reflects the measurements with a high accuracy. The largest errors of the fits occurred with measurements in wall region of the fluidized bed at larger superficial gas velocities and distance from the gas distributor (figure 5.19 (d) and (f)). The number of detected bubbles per time is much lower at the wall under these conditions in comparison to the center of the bed. The reason for this are the preferred bubble flow ways in the center of the fluidized bed, whereas the back-flow of solids mainly takes place in the wall region. A low amount of bubbles detected increases the error of measurement because the statistical reliability decreases significantly.

In figure 5.19 (a) more than 50 % of the bubbles near the wall and in the center of the bed have a pierced length smaller than 2 cm. With increase of the superficial gas velocity larger bubbles occur but small bubbles still do exist. A broader distribution of pierced lengths is the result. Furthermore, a lateral distribution of bubble sizes forms. In the wall region small bubbles are more likely to be formed in comparison to the fluidized bed center as can be seen in figures 5.19 (c) and (e). This behavior happens due to the occurrence of preferred bubble flow paths resulting from coalescence. Furthermore, the back-flow of solids at the fluidized bed wall increases the resistance for bubbles rising in this region of the bed. In order to minimize the drag bubbles move to the center of the bed, especially when they become large.

Vertical bubble sizes larger than the fluidized bed diameter ($D - 0.1$ m) can be measured at a superficial gas velocity of $0.7\,\mathrm{m\,s^{-1}}$ and larger. This leads to the conclusion that bubbles at these sizes must have a lower lateral than vertical elongation. These findings contrast the assumption of spherical cap shaped bubbles as they are described in literature [59].

With increasing distance from the gas distributor bubble sizes increase as can be seen in the figures 5.19 (a), (c) and (e) in comparison to (b), (d) and (f), respectively. The probability of the occurrence of small bubbles decreases due to coalescence phenomena. The distributions increase and a broader spectrum of pierced lengths results.

By comparison of the different fluidized bed diameters investigated it can be found that in larger fluidized beds broader distributions of the pierced lengths are measured. The fitted probability density distributions (without the histograms) for similar measurement conditions are shown in figure 5.20. The broader

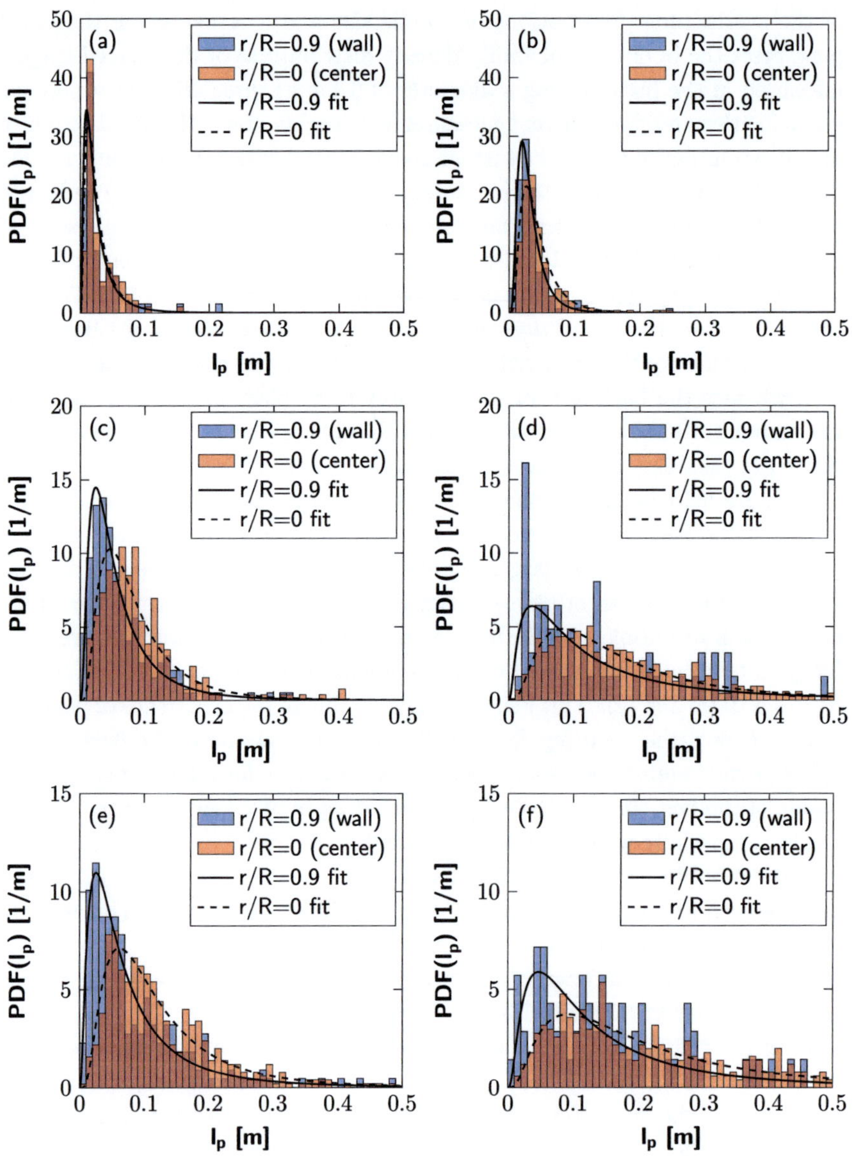

Fig. 5.19: Probability density functions of pierced lengths for $D = 0.1\,\mathrm{m}$ and $H_0 = 0.2\,\mathrm{m}$ (sand B1) at: (a) $z = 0.05\,\mathrm{m}$, $U_0 = 0.18\,\mathrm{m\,s^{-1}}$, (b) $z = 0.15\,\mathrm{m}$, $U_0 = 0.18\,\mathrm{m\,s^{-1}}$, (c) $z = 0.05\,\mathrm{m}$, $U_0 = 0.7\,\mathrm{m\,s^{-1}}$, (d) $z = 0.15\,\mathrm{m}$, $U_0 = 0.7\,\mathrm{m\,s^{-1}}$, (e) $z = 0.05\,\mathrm{m}$, $U_0 = 1.4\,\mathrm{m\,s^{-1}}$, and (f) $z = 0.15\,\mathrm{m}$, $U_0 = 1.4\,\mathrm{m\,s^{-1}}$.

distributions measured lead to larger arithmetic means of the distributions. One reason for the differences in the distributions is a higher probability of coalescence in larger fluidized bed facilities. In a fluidized bed having a small diameter the lateral size of bubbles reaches the magnitude of the bed already at low heights above the gas distributor and low superficial gas velocities. This makes lateral coalescence impossible. Furthermore, these bubbles have a large resistance due to the back-flow of solids at the wall which limits the rise velocity of these bubbles. In addition, this decreases the probability of vertical coalescence. The fluidized bed reaches a semi-slugging state where bubble sizes do not completely reach the diameter of the bed and bubble size is limited. Another reason for broader bubble size distributions in larger fluidized beds is the formation of larger bubbles at the gas distributor [13]. Areas of deprived bubble flow and solid back-flow are larger with increasing size of a fluidized bed. In contrast to a small fluidized bed, where the stability of large bubbles is lower due to the direct interaction of back-flowing solids with the bubble in the generation process, the pressure gradient in a large plant responsible for bubble formation acts at a larger cross-sectional area. Thus, large bubbles in large beds can be formed more easily.

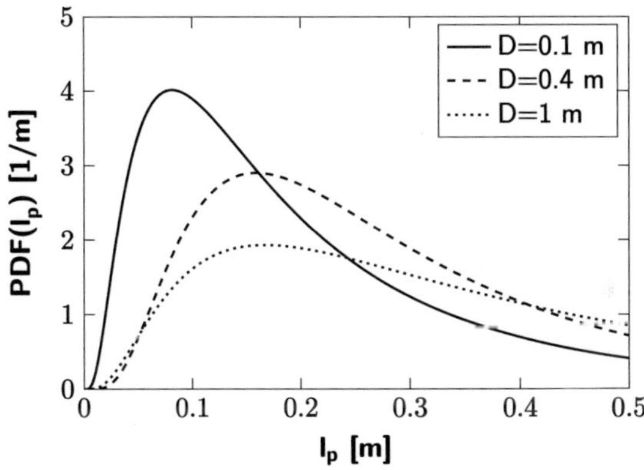

Fig. 5.20: Pierced length in the center of the fluidized bed for sand B1 at different plant diameters ($D = 0.1\,\mathrm{m}$ measured at $H_0 = 0.3\,\mathrm{m}$, $z = 0.2\,\mathrm{m}$; $D = 0.4\,\mathrm{m}$ measured at $H_0 = 0.4\,\mathrm{m}$, $z = 0.19\,\mathrm{m}$, $D = 1\,\mathrm{m}$ measured at $H_0 = 1\,\mathrm{m}$, $z = 0.18\,\mathrm{m}$) and superficial gas velocities of $U_0 = 1.1\,\mathrm{m\,s^{-1}}$ (for $D = 0.1\,\mathrm{m}$ and $D = 1\,\mathrm{m}$), and $U_0 = 1\,\mathrm{m\,s^{-1}}$ (for $D = 0.4\,\mathrm{m}$).

For comparison of different fluidizing conditions the arithmetic mean values of the fitted logarithmic normal distributions according to equation 3.18 are used. Furthermore, these values are averaged over the cross-sectional area of the fluidized bed under consideration of the amount of bubbles occurred per time.

The result is the cross-sectional averaged mean pierced length that is discussed in the following.

The dependencies of the cross-sectionally averaged mean pierced length from superficial gas velocity and distance from the gas distributor can be seen in figure 5.21 (a) for a plant diameter of 0.1 m and (b) a plant diameter of 1 m. In turbulent fluidization as well as in the bubbling regime the vertical lengths of bubbles are found to increase with superficial gas velocity. In literature the maximum of pressure fluctuations, which defines the entrance from bubbling to turbulent fluidization, is described to happen due to the reduction of bubble sizes because of an increasing rate of bubble break-up [6]. If bubbles are assumed to be in the characteristic spherical cap shape as described for the bubbling regime in literature [59, 60], the assumption of a decreasing bubble size would lead to smaller pierced lengths, which stands in contrast to the results found in this work. But as explained before, due to the large vertical sizes of bubbles in comparison to the fluidized bed size the assumption of a spherical cap is not valid at velocities above the low velocity bubbling state (velocities slightly above minimum fluidization velocity).

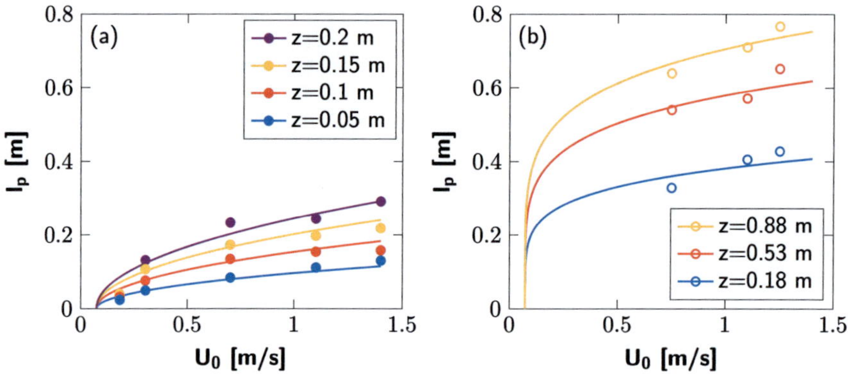

Fig. 5.21: Cross-sectionally averaged mean pierced lengths measured at different heights above the gas distributor and their fits (solid lines) according to equation 5.13 in dependence of superficial gas velocity at (a) D=0.1 m and (b) D=1 m (sand B1) acc. to [123].

With increasing distance from the gas distributor the cross-sectionally averaged mean pierced length also increases as shown in figure 5.21. This behavior is the result of broader size distributions with increasing distance from the gas distributor that were described before with figure 5.19. Equation 5.13 gives an empiric approach for the parameter dependencies (bed diameter, superficial gas velocity and distance from gas distributor) of the cross-sectionally averaged mean pierced length and is a fit to the measurement data for sand B1 in the fluidized bed plants having diameters of 0.1 m, 0.4 m and 1 m (all parameters in SI units, cf. [123]):

$$l_p = 0.751 z^{0.356 \frac{D+0.089}{D}} (U_0 - U_{mf})^{0.148 \frac{D+0.232}{D}} \tag{5.13}$$

Predictions of the mean pierced length by this equation are shown as solid lines in the same color code as the measured data points in figure 5.21. Equation 5.13 slightly overestimates the cross-sectionally averaged mean pierced length at a superficial gas velocity of $0.18\,\mathrm{m\,s^{-1}}$ as shown in figure 5.21 (a). At this superficial gas velocity bubbles are small in comparison to the fluidized bed plant and the formation of a spherical cap shape is more probable. The parity graph in figure 5.22 shows an error below 10 % for equation 5.13 in most cases.

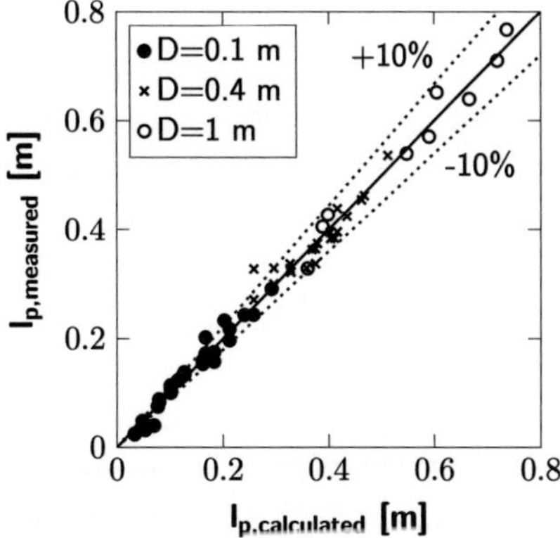

Fig. 5.22: Parity graph of measured and fitted cross-sectionally averaged mean pierced length for all plant diameters investigated acc. to [123].

The measurements show that bubble growth in vertical direction is unlimited for the kind of particles investigated. This agrees with observations from literature, where unlimited bubble growth for particles of Geldart's group B is predicted [108, 110]. Because particles belonging to this group according to the Geldart classification have similar fluidization behavior, it is assumed that equation 5.13 can be used for the prediction of the cross-sectionally averaged mean pierced length of different materials of this group directly or by small variation.

5.3.2 Bubble Velocity

In literature most studies measuring the velocity of bubbles focus on the bubbling regime at superficial gas velocities below $0.3\,\mathrm{m\,s^{-1}}$ [13]. To get information about bubble velocities in the turbulent regime a broad range of superficial gas velocities from $0.18\,\mathrm{m\,s^{-1}}$ to $1.6\,\mathrm{m\,s^{-1}}$ was investigated in this study.

The rise velocity of a bubble in a fluidized bed depends on its size and the influence of swarm behavior of multiple bubbles rising (see section 2.2.2). Due to the variety of bubble sizes caused by generation processes and coalescence a distribution of different rise velocities of bubbles is always measured at one measurement point. The probability density distributions of the bubble rise velocities at the same conditions described in the graphs in figure 5.19 are shown in figure 5.23, respectively. The distribution of bubble velocities behaves similar to the distributions of the pierced lengths measured. At a superficial gas velocity of $0.18\,\mathrm{m\,s^{-1}}$ in the bubbling regime near the gas distributor (figure 5.23 (a)) most of the bubbles have velocities in a small range, whereas with larger superficial gas velocity the distribution broadens and larger bubble velocities are measured (figure 5.23 (c) and (e)). The pierced lengths measured at these superficial gas velocities also increase because of the effects described in section 5.3.1. Thus, this behavior agrees with observations in literature that bubble velocity is strictly dependent on the size of a bubble. As it can be seen in the graphs (b), (d) and (f) in comparison to (a), (c) and (e) in figure 5.23 the distribution measured of bubbles velocities broadens with increasing distance from the gas distributor due to the growth of bubbles and the resulting larger rise velocities.

The effects of the radial measurement position on the bubble velocity distributions are the same as for the pierced length distributions. Due to larger bubbles occurring preferably in the fluidized bed center the rise velocity of these bubbles increases as well. Because of the small amount of bubbles detected at the wall in the graphs (d) and (f) in figure 5.23 the error of the distributions measured is larger than at other positions. Due to the small fixed distance of the capacitance probe channels to each other and the recording frequency of $10\,\mathrm{kHz}$ the histograms show lags at larger velocities. Logarithmic normal distributions are fitted to the histograms, which give an approach to the measured distributions with a high accuracy in cases where significant amounts of bubbles are determined and evaluated.

Figure 5.24 shows the fitted probability density distributions of the bubble velocities at different fluidized bed diameters and similar operation conditions. Analogous to the pierced length the distribution of bubble velocities broadens with increasing bed diameter. This happens due to the direct link of bubble velocity to the pierced length.

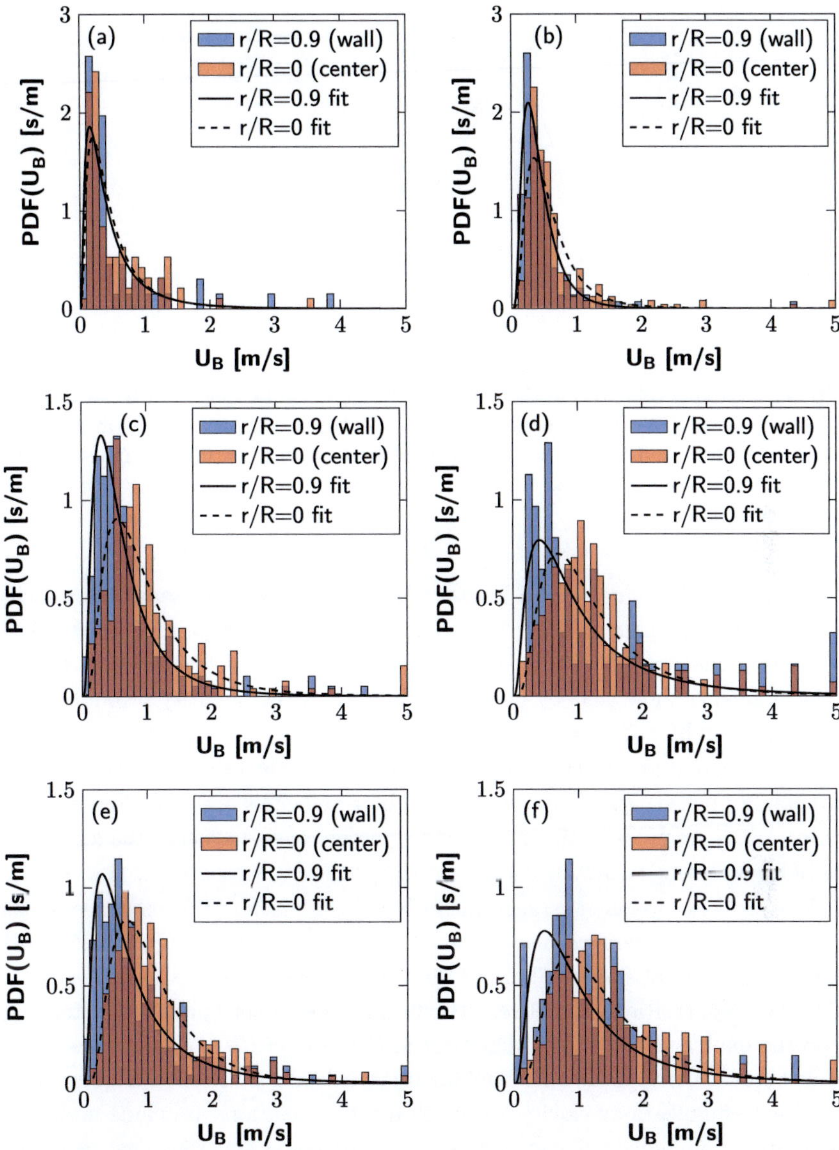

Fig. 5.23: Probability density functions of bubble velocities for $D = 0.1\,\mathrm{m}$ and $H_0 = 0.2\,\mathrm{m}$ (sand B1) at: (a) $z = 0.05\,\mathrm{m}$, $U_0 = 0.18\,\mathrm{m\,s^{-1}}$, (b) $z = 0.15\,\mathrm{m}$, $U_0 = 0.18\,\mathrm{m\,s^{-1}}$, (c) $z = 0.05\,\mathrm{m}$, $U_0 = 0.7\,\mathrm{m\,s^{-1}}$, (d) $z = 0.15\,\mathrm{m}$, $U_0 = 0.7\,\mathrm{m\,s^{-1}}$, (e) $z = 0.05\,\mathrm{m}$, $U_0 = 1.4\,\mathrm{m\,s^{-1}}$, and (f) $z = 0.15\,\mathrm{m}$, $U_0 = 1.4\,\mathrm{m\,s^{-1}}$.

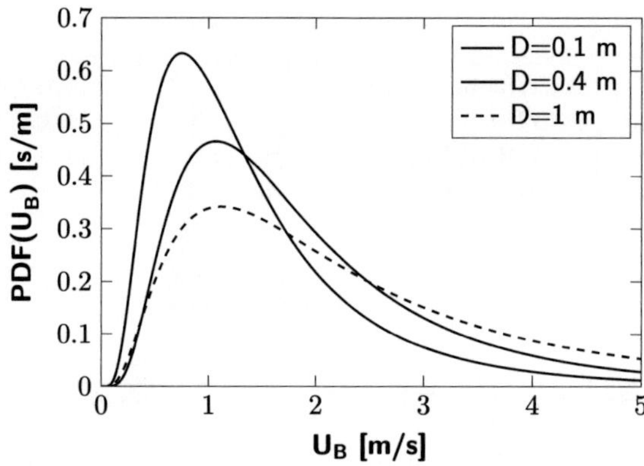

Fig. 5.24: Bubble velocities in the center of the fluidized bed for sand B1 at different plant diameters ($D = 0.1$ m measured at $H_0 = 0.3$ m, $z = 0.2$ m; $D = 0.4$ m measured at $H_0 = 0.4$ m, $z = 0.19$ m, $D = 1$ m measured at $H_0 = 1$ m, $z = 0.18$ m) and superficial gas velocities of $U_0 = 1.1\,\mathrm{m\,s^{-1}}$ (for $D = 0.1$ m and $D = 1$ m), and $U_0 = 1\,\mathrm{m\,s^{-1}}$ (for $D = 0.4$ m).

Bubbles of the same pierced length do not always have the same rise velocity. Their velocity is influenced by their swarm behavior, location, and shape, which leads to a distribution of bubble velocities for one pierced length. Nevertheless, measurements show that by determination of a large quantity of bubbles the velocities can be allocated to a mean pierced length.

To compare different fluidizing conditions the arithmetic mean values of the probability density distributions of the bubble velocities are used. Furthermore, these values are cross-sectionally averaged under consideration of the amount of bubbles occurring within measurement time at each measurement point.

The cross-sectionally averaged mean bubble velocities in dependence on the cross-sectionally averaged mean pierced length are shown in figure 5.25. With increasing vertical size of a bubble the velocity of a bubble also increases. In literature the relationship between the volume equivalent bubble diameter $d_{B,V}$ and the bubble velocity U_B is described in the form of $U_B \sim d_{B,V}{}^{0.5}$ mostly [13]. This approach is applicable for bubbles at superficial gas velocities slightly above minimum fluidization velocity for which most correlations are made under the assumption of a spherical cap bubble shape. At larger superficial gas velocities, especially in turbulent fluidization, this approach leads to an overestimation of the bubble velocity, which was also found in literature [6, 108]. According to figure 5.25 the bubble velocity is linearly dependent on the pierced length. Thus, equation 5.14 is proposed to calculate the bubble velocity U_B from the pierced length l_p (all parameters in SI units, cf. [123]):

$$U_B = 0.587\frac{U_0 - U_{mf}}{U_0} + 4.344 l_p \tag{5.14}$$

This empiric equation is fitted to the measurements with sand B1 and considers the influence of swarm behavior with increasing excess gas velocity $U_0 - U_{mf}$ and superficial gas velocity U_0. The influence of the swarm behavior in this case increases with increasing superficial gas velocity. This effect comes from more bubble interactions due to a higher bubble density in fluidized beds at larger superficial gas velocities (see bubble phase holdup in section 5.2). At velocities above circa $1\,\mathrm{m\,s^{-1}}$ the superficial gas velocity is large in comparison to the minimum fluidization velocity ($U_0 \gg U_{mf}$). This simplifies equation 5.14 to the solid line plotted in figure 5.25. At velocities fulfilling this condition the influence of swarm behavior is limited and reaches a maximum. The bubbles reach large vertical sizes and the turbulent fluidized bed state is reached. This results in different bubble interaction behavior than in the bubbling regime.

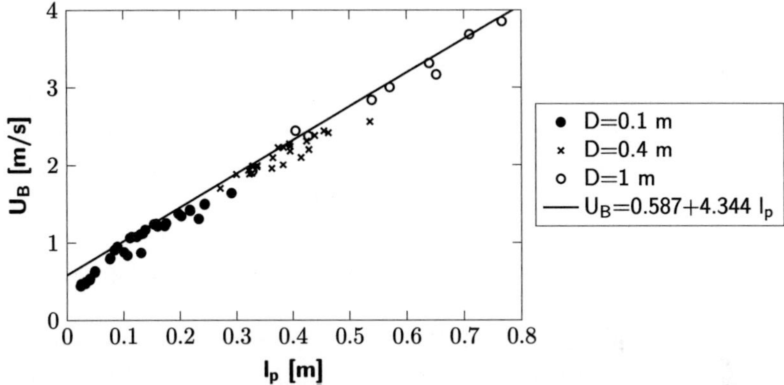

Fig. 5.25: Bubble velocity in dependence on the mean pierced length at different plant diameters, heights above the gas distributor, and superficial gas velocities acc. to [123] (cross-sectionally averaged, sand B1).

The parity graph of equation 5.14 and relating experimental data are shown in figure 5.26. A maximum error of 12 % occurs for measurements at positions high above the gas distributor in comparison to the fluidized bed diameter ($z/D > 1.5$) at superficial gas velocities in the bubbling fluidized bed regime. At these heights bubbles become large in comparison to the bed diameter and get influenced by the back-flow of solids. The attempt of a bubble to create a spherical cap like shape works against the wall influence and the down coming solids. This results in larger drag forces acting on the bubbles and thus a reduction of the rise velocity which can be observed.

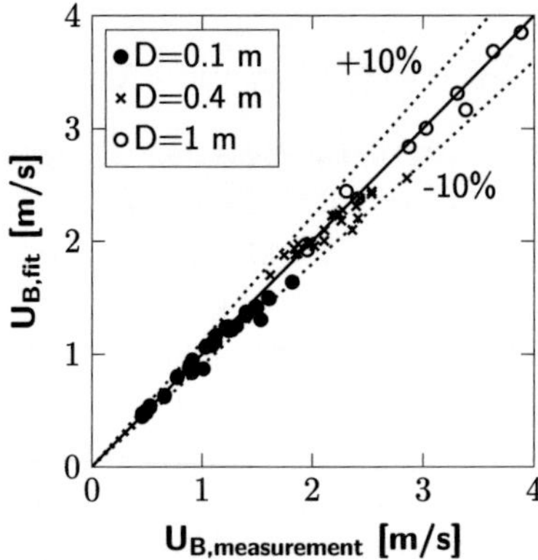

Fig. 5.26: Parity graph of measured bubble velocities and fitted ones in dependence on the mean pierced length of the bubbles acc. to [123].

5.3.3 Bubble Shape

Fluidized beds are systems that attempt to minimize energies. Above minimum fluidization velocity a large fraction of the gas rises as bubbles in order to reduce the drag acting on the gas. In contrast to gas bubbles in liquids, bubbles in fluidized beds do not have a surface tension, which has a large influence on the bubble shape in gas-liquid systems [59, 60]. Nevertheless, similarities between gas-liquid and gas-solid systems were found. Spherical cap bubbles in fluidized beds are assumed to behave like large bubbles in high viscous liquids having a drag coefficient of 2.64 [11].

The measurements of the pierced length shows that bubbles must change their shape due to the influence of the fluidized bed diameter at bubbling fluidization in small plants and due to the occurrence of bubbles having a large vertical size in comparison to the bed diameter in turbulent fluidization. In the first case bubbles are compressed laterally because of the reach of a semi-slugging state as explained before (section 5.3.1). In the case of turbulent fluidization vertical bubble size grows further with increasing superficial gas velocity, whereas pressure fluctuation intensity decreases. Spherical cap shaped bubbles of the sizes measured at superficial gas velocities in the turbulent regime would lead to even larger pressure fluctuations due to their large pressure gradient and the large eruption at the bed surface. This leads to the conclusion that bubbles at large velocities must have a different shape that is smaller in lateral direction than a spherical cap shaped bubble having the same vertical length.

Because capacitance probes only allow the measurement of vertical bubble properties the flow structure in the fluidized bed plant having a diameter of

0.1 m (made of acrylic glass) was investigated visually with a high speed camera. Two frames of the recordings can be found in figure 5.27 in (a) bubbling and (b) turbulent regime. Due to the fact that the fluidized bed facility is three dimensional and the camera visualizes particles and bubbles that directly move at the wall, only qualitative information about the bubble shape can be obtained. Furthermore, the preferred bubble flow paths in the center of the bed cause bubbles to move at the wall at rare intervals. For better visualization the outlines of the bubbles are contoured in black in the figure. A clear difference of bubble shapes in bubbling and turbulent fluidization is observed. Bubbles at a velocity of $0.21\,\mathrm{m\,s^{-1}}$ in the bubbling regime are found to rise in a spherical shape pulling a wake behind as figure 5.27 (a) shows. In contrast to this, in turbulent fluidization (see figure 5.27 (b)) bubbles do not rise in the typical spherical cap shape anymore. The bubbles detected have a larger vertical dimension than laterally at the wall as suggested by capacitance probe measurements.

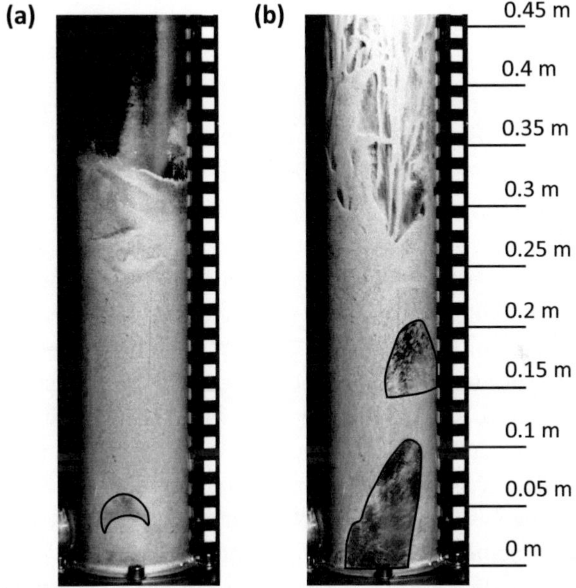

Fig. 5.27: Photographs of a fluidized bed ($D = 0.1\,\mathrm{m}$, $H_0 = 0.2\,\mathrm{m}$, sand B1) in the (a) bubbling ($U_0 = 0.21\,\mathrm{m\,s^{-1}}$) and (b) turbulent ($U_0 = 1.5\,\mathrm{m\,s^{-1}}$) state recorded with a high-speed camera and with marked bubble outlines.

The drag coefficient of spheres and bubbles in liquids directly depends on the Reynolds number [206]. The characteristic length for calculating the Reynolds number in the case of a rising bubble is its lateral diameter assuming rotational symmetry. Thus, the drag coefficient of a bubble in liquids directly depends on its lateral size. In fluidized beds further parameters like the wake formation and swarm behavior do influence the drag coefficient of a bubble. But it is obvious that the lateral size of a bubble has an influence on the drag coefficient that is not negligible. Using equation 2.9, which is based on a force balance of drag force, buoyancy force and gravitational force, the drag coefficient of a bubble in a fluidized bed can be calculated from its pierced length and its velocity under consideration of the following assumptions:

- Bubbles are ellipsoids of revolution having a lateral diameter and a vertical diameter (pierced length);

- The density of the fluidized bed ρ_{FB} is much larger than the density of the fluidizing fluid ρ_f ($\rho_{FB} \gg \rho_f$);

- Swarm behavior of bubbles in the fluidized bed does not influence the drag of a single bubble.

The result is

$$C_D = \frac{4}{3} g \frac{l_p}{U_B{}^2},\qquad(5.15)$$

which describes the drag coefficient C_D in direct relation to pierced length l_p, bubble velocity U_B and gravitational acceleration g.

The results for the drag coefficients calculated using the cross-sectionally averaged mean pierced lengths and cross-sectionally averaged mean bubble velocities measured for the three different bed diameters investigated are shown in figure 5.28. In addition to this, the predicted drag coefficients are calculated by the combination of equation 5.15 with equation 5.14. They are plotted as curves for different superficial gas velocities. As it can be seen in the figure the calculated drag coefficients first increase with the increasing pierced length, reach a maximum at pierced lengths around $0.08 - 0.13\,\text{m}$ and finally decrease at further growth. Bubbles with a larger lateral size might have larger wakes than smaller ones (see section 2.2.5.2). Furthermore, these bubbles have to displace larger amounts of solids and the increasing buoyancy forces them to rise faster. These effects lead to an increase of the drag coefficients that is observed in the first section described in figure 5.28 where the drag coefficient increases with increasing pierced length. Bubbles in this section are observed to mainly occur in the bubbling state of the fluidized bed. These bubbles grow with increasing vertical size also in lateral size. The shape of these bubbles can be close to the spherical cap shape described in literature, whereas small bubbles can have smaller wakes than larger ones [60] due to the high inertia of the particles especially with particles of Geldart's group B. Larger bubbles at sizes where the drag coefficient decreases with increasing vertical size change their shape to reduce drag. The lateral size of these bubbles decreases and a decrease of the wake fraction of the bubble is the result. The shape of lowest resistance for these bubbles is a vertically stretched shape (pierced length is larger than lateral size). Bubbles in the turbulent fluidized bed regime generally have drag coefficients smaller than at bubbling fluidization according to figure 5.28. These bubbles moving at large velocities might never form a spherical cap shape and rise directly as vertical stretched bubbles. This agrees with the visual observations found in figure 5.27.

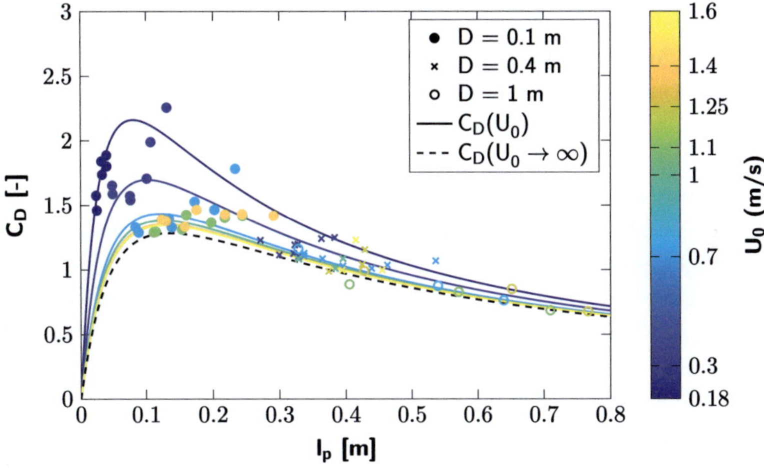

Fig. 5.28: Bubble drag coefficient in dependence of the bubble mean pierced length acc. to [123].

5.3.4 Effect of Bubble Shape on the Transition from Bubbling to Turbulent Fluidization

The transition from bubbling to turbulent fluidization is defined by a maximum in the amplitude of the pressure fluctuations of a fluidized bed as discussed in detail in chapter 4. These pressure fluctuations are mainly caused by bubble formation, coalescence, bubble break-up and bubble eruption at the surface. If a laterally large bubble rises in a fluidized bed, it displaces and carries more solids in cloud and wake than a smaller one. The pressure gradients induced by this bubble are generally larger. When the bubble erupts at the surface large amounts of solids are carried into the freeboard. This leads to a growth of pressure fluctuation intensity with increasing lateral bubble size.

As explained in section 5.3.3 bubbles must decrease in lateral size at large superficial gas velocities. Bubbles that are stretched in vertical dimension have smaller wakes and clouds than bubbles having a spherical cap shape. They carry less solid material per bubble volume. Thus, the influence of these bubbles on pressure fluctuations is lower. Bubbles change their shape with increasing superficial gas velocity to an elongated form, whereas at the same time the amount of bubbles rising in the fluidized bed increases. A maximum in pressure fluctuation intensity can be found at the point where the reduction of the bubble volume specific forces overcomes the influence of the increasing number of bubbles. Superficial gas velocities beyond this point lead to a further decrease in lateral bubble size. In addition to this, particle entrainment begins and solid material carried into freeboard due to bubble eruption does not fall back immediately. Thus, the typical flow structure of the turbulent regime is reached with high heat and mass transfer abilities due to the mixing behavior of the thin bubbles rising.

In literature a downwards migration of the maximum in pressure fluctuation amplitudes and thus the transition from bubbling to turbulent fluidization with increasing superficial gas velocity is found [6, 119, 120]. This happens due to different vertical bubble sizes occurring at different heights above the gas distributor. Bubbles in larger heights are larger and they have to change their shape at lower superficial gas velocities to reduce drag in contrast to bubbles closer to the gas distributor.

Furthermore, a later transition from bubbling to turbulent fluidization can be found at the same static bed heights in beds of smaller diameters [6, 101, 104, 112, 113]. In small fluidized bed plants bubbles are forced to coalesce more, even at larger superficial gas velocities in comparison to larger plants, because of preferred flow ways of gas and solids as explained before. Thus, to reduce the lateral size of bubbles rising in small fluidized beds at the same conditions as in beds of larger diameters, higher superficial gas velocities are necessary.

5.4 Gas Short-Circuiting

The occurrence of gas short-circuiting in fluidized beds as described in section 2.2.6 is a major issue for the prediction of the reaction behavior of a fluidized bed. Gas passing the fluidized bed at a larger velocity than the bubble rise velocity or interstitial gas velocity has a lower residence time in the fluidized bed. Thus, gas-solids contact time is reduced leading to lower yields in chemical conversion.

While the visible volume flow of gas passing the fluidized bed in form of bubbles can be determined by several measurement techniques, the determination of the volume flow of short-circuiting gas is challenging. To calculate this volume flow, knowledge about the gas flow in the suspension phase (interstitial gas velocity) is indispensable (see eq. 2.13). As the measurements showed, the expansion of the suspension phase increases at larger superficial gas velocities. For this reason, the interstitial gas velocity is assumed to be larger than the minimum fluidization velocity. To estimate the interstitial gas velocity in dependence on the void fraction of the suspension phase the Ergun equation (eq. 2.1) is used under the assumption that the ratio of pressure drop and height difference is the same for an expanded bed of particles and a bed at minimum fluidization velocity. For the calculation of the interstitial gas velocity U_{int} in dependence on the bed porosity ϵ

$$U_{int} = -N + \sqrt{N^2 + \frac{150}{1.75} \frac{(1 - \epsilon_{mf})\,\epsilon^3}{\epsilon_{mf}^3} \frac{\eta_f}{\rho_f d_p} U_{mf} + \frac{\epsilon^3}{\epsilon_{mf}^3} U_{mf}^2} \qquad (5.16)$$

$$N = \frac{75}{1.75} (1 - \epsilon) \frac{\eta_f}{\rho_f d_p}$$

results. According to this approach, the interstitial gas velocity depends on the bed porosity at minimum fluidization ϵ_{mf}, the minimum fluidization velocity U_{mf}, gas density ρ_f, gas viscosity η_f and particle size d_p.

By estimation of the interstitial gas velocity using equation 5.16 and determination of the volume flow of gas in the bubble phase by capacitance probe the short-circuiting volume flow and the short-circuiting velocity U_{sc} (related to the cross-sectional area of the bed) can be calculated. This velocity is shown in figure 5.29 (a) in dependence of the distance from the gas distributor at different superficial gas velocities. The gas short-circuiting velocity generally increases with increasing superficial gas velocity. At larger superficial gas velocities more bubbles are generated at the gas distributor and the bubble density in the fluidized bed increases. This means a lowering of the average bubble distance between each other leading to a more simplified short-circuiting process between bubbles as phenomenological described in literature [75]. Shen *et al.* [65] describes a linear increase of the gas short-circuiting with increasing superficial gas velocity. Similar behavior can also be found in figure 5.29 (b). Thereby, no dependence on the bed diameter can be determined. Furthermore, in 5.29 (a) a decrease of the gas short-circuiting can be found with increasing distance from the gas distributor, as also reported in literature [65]. Bubbles coalesce during their rise in the fluidized bed and increase in rise velocity. This leads to larger bubbles occurring at larger distances to each other. For this reason, if gas short-circuiting mainly occurs between bubbles close to each other, it is impeded in the upper regions of the bed.

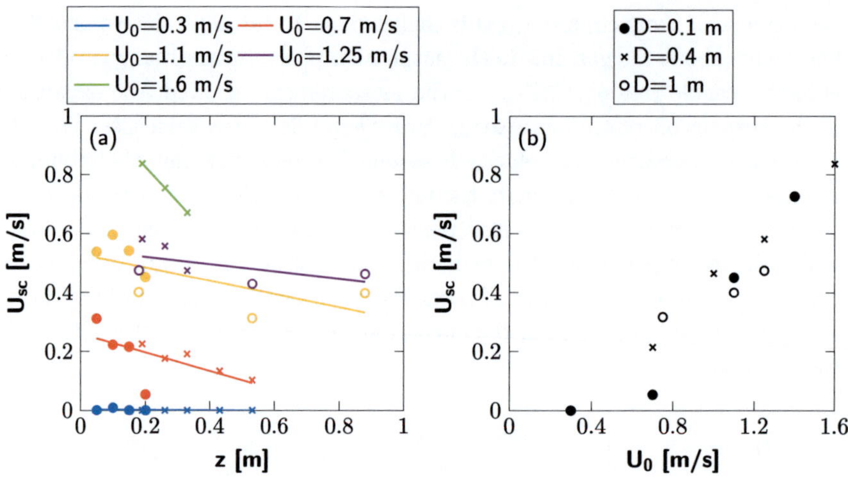

Fig. 5.29: Short circuiting velocity U_{sc} for sand B1 in dependence of (a) the height above the gas distributor and (b) the superficial gas velocity at in fluidized bed plants of different sizes.

Figure 5.30 shows the volume flow ratios of suspension phase, visible bubble phase and gas short-circuiting in dependence of the superficial gas velocity. The ratio of gas going through the suspension phase steadily decreases at larger superficial gas velocities. In contrast to this, in the bubbling regime the ratio of short-circuiting increases sharply. At a superficial gas velocity of about $1\,\mathrm{m\,s^{-1}}$ almost half of the gas passes through the bed by short-circuiting. Above this superficial gas velocity in the turbulent regime ($U_c = 0.94\,\mathrm{m\,s^{-1}}$) the ratio of the gas short-circuiting volume flow flattens. The visible bubble volume flow ratio shows inverse behavior. In the turbulent regime an equilibrium between the flow ratios seems to set in. Due to the large amounts of gas going through the fluidized bed in this state the flow behavior changes and gas short-circuiting does not gain further significance.

5.5 Entrainment

The rate of entrainment from the fluidized bed is determined in the standpipe of the CFB400 as explained in section 3.3.3. As the results in figure 5.31 show, the solids circulation rate G_s was found to be independent of the static bed height. The error of the measurements are small and the results have a high reproducability. The rate of entrainment steadily increases at larger superficial gas velocities up into the fast fluidization regime. The correlation given by

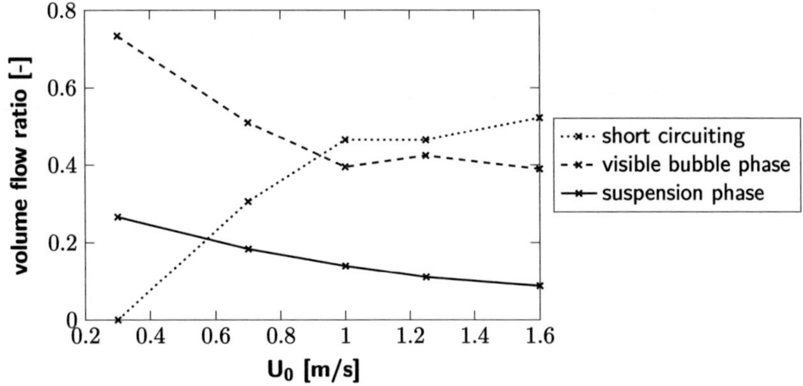

Fig. 5.30: Volume flow ratios of gas going through the bed of sand B1 in the suspension phase, the visible bubble phase and by short circuiting at $D = 0.4\,\text{m}$, $H_0 = 0.4\,\text{m}$ and $z = 0.19\,\text{m}$ ($U_c = 0.94\,\text{m\,s}^{-1}$).

Tasirin and Geldart [149], which is usually used for the determination of the entrainment from turbulent fluidized beds according to Fotovat [147], highly overestimates the measured rates of entrainment in figure 5.31. This might be related to the reason that this correlation was determined for particles of Geldart's group A. The entrainment measurements carried out in this work have been done for sand B1, which belongs to Geldart's group B. The differences in the fluidization behavior of both groups might also lead to different entrainment behavior.

Other correlations given by Choi $et\ al.$ [207], Geldart $et\ al.$ (acc. to [208]) or listed by Werther and Hartge [148] give insufficient data predictions of the entrainment rate as well. For this reason, a new correlation for the elutriation rate based on the entrainment measurements of this work and the correlative approach of Tasirin and Geldart [149] (see eq. 2.45) is introduced by the following equation:

$$K_i^* = 16.7\rho_f U_0^{1.7} exp\left(-5.4\frac{U_{t,i}}{U_0}\right), \tag{5.17}$$

with all parameters in SI units. The fit predicts the rates of elutriation and entrainment measured in this work with high accuracy ($R_{corr}^{\,2} = 0.96$).

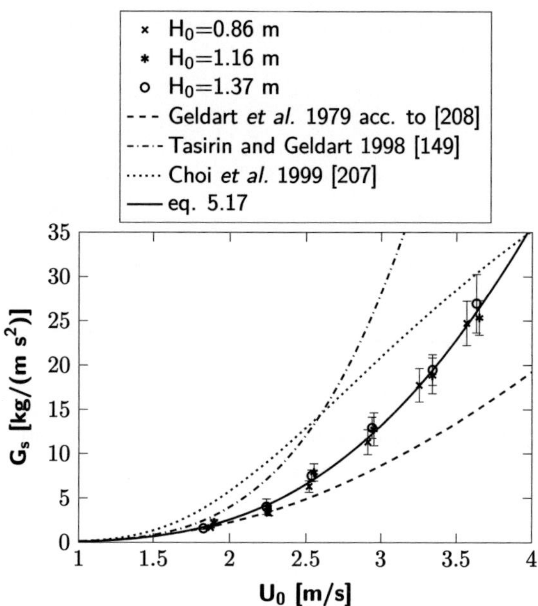

Fig. 5.31: Measured solids circulation rates for sand B1 compared to calculated entrainment rates from literature and by equation 5.17.

6 Fluid Dynamic Model

Based on the findings of the chapters 4 and 5 a model describing the fluid dynamic properties of turbulent fluidized beds is introduced in this chapter. The model predictions of solids concentrations and pressures are validated with measurement data from fluidized bed experiments.

6.1 Modeling Approach

The model proposed is based on correlations derived from the measurements carried out in the facilities of different diameters described in chapter 3. In addition to this, for the determination of certain parameters approaches from literature are applied.

6.1.1 Axial Solids Concentration

The model approach distinguishes between two major zones of the fluidized bed: a dense zone which contains most of the solid material and a freeboard zone where particles get entrained. The dense zone is assumed to consist of two phases, a suspension phase and a bubble (or void) phase. Thus, the two-phase theory from literature is followed. Thereby, the bubble phase is assumed to be empty of solids. The movement of particles in the dense zone is described by an acceleration and rise in the bed center induced by the rising bubbles. In contrast to this, at the fluidized bed wall a region of solids downflow occurs. In both regions of the dense zone the occurrence of high solids concentrations is possible with an increased probability at the fluidized bed wall. Despite this flow behavior, the model proposal is one-dimensional. A prediction of the radial change of concentrations is not considered. All solids concentrations predicted by the model are averaged over the fluidized bed cross-section.

Above the dense zone, in the freeboard zone of the fluidized bed the solids concentration decreases and particles are decelerated. Depending on the superficial gas velocity, particles get entrained and a dilute rise of solids is the result.

The height h depending cross-sectional averaged solids concentration $c_V(h)$ is described by a logistic function:

$$\frac{c_V\left(h\right) - c_{V,\infty}}{c_{V,D} - c_{V,\infty}} = 1 - \left(1 + exp\left(-a - a\left(h - H_{exp}\right)\right)\right)^{-exp(a)} \tag{6.1}$$

Here, $c_{V,\infty}$ is the solids concentration above the transport disengaging height (TDH), $c_{V,D}$ the solids concentration in the dense zone, a the exponential decay constant and H_{exp} the expansion height of the bed marking the point of transition from the dense to the freeboard zone. The advantage of using a logistic function in contrast to an immediate, sharp transition from dense zone to freeboard as proposed by Werther and Wein [110] is the smooth transition between both zones describing this region more sufficient.

The solids concentration $c_{V,\infty}$ results from the amount of entrained solids. Under the assumption of the particles rising above TDH with the difference of superficial gas velocity U_0 and mean terminal velocity of the particles U_t, this concentration can be calculated from the rate of entrainment G_s and the particle density ρ_p:

$$c_{V,\infty} = \frac{G_s}{\left(U_0 - U_t\right)\rho_p} \tag{6.2}$$

For the determination of the entrainment rate correlations are available in literature (see section 5.5). Due to the entrainment of the material used in this work being described insufficiently by literature approaches, equation 5.17 (elutriation) in combination with equation 2.46 is used here. The terminal velocity of the particles can be calculated as described in section 3.2.4 by iteration of the equations 3.5, 3.6 and 3.7.

The solids concentration of the dense phase $c_{V,D}$ is determined by

$$c_{V,D} = \left(1 - \phi_B\right)\left(\psi_{S,low}c_{V,S,low} + \left(1 - \psi_{S,low}\right)c_{V,S,high}\right) \tag{6.3}$$

using equation 5.1 to calculate the solids concentration of the high concentration suspension phase $c_{V,S,high}$, equation 5.2 to determine the solids concentration of the low concentration suspension phase $c_{V,S,low}$, equation 5.9 to estimate the bubble holdup ϕ_B and equation 5.11 to determine the holdup of the low concentration suspension phase $\psi_{S,low}$.

The exponential decay constant a can be calculated from literature approaches as listed in table 2.2. To describe the decay for the sand used in this work the approach given by equation 2.31 was found to be most suitable.

6.1.2 Mass Balances

For determination of the expanded bed height H_{exp} mass balances must be solved. The mass balance of the fluidized bed (FB) is given by

$$m_{FB} = \int_0^{H_{FB}} A_0 c_V\left(h\right) \rho_p dh \tag{6.4}$$

The parameter H_{FB} describes the whole height of the fluidized bed reaching from the gas distributor up to the exit.

Due to larger rates of entrainment with increasing superficial gas velocity, larger amounts of solids remain in the other parts of the facility such as the cyclone, the standpipe (SP) and the loop-seal (LS). This leads to a reduction of the amount of bed material in the fluidized bed itself resulting in different fluidization behavior. For this reason, the amount of material remaining in the other plant elements must be approximated.

With increasing superficial gas velocity the pressure drop in the upper regions of the fluidized bed increases. This happens due to larger amounts of bed material remaining at larger heights. Below the standpipe a loop-seal is usually installed to guarantee that fluidization gas does not bypass through the standpipe. Above the loop-seal entrained solids coming from the cyclone accumulate and form a moving bed. The height of this bed is directly related to the pressure drop induced by it. The latter equals the pressure drop of the fluidized bed from the solids return (SR) up to the top $\Delta P_{H_{SR}-H_{FB}}$ (assumption: cyclone pressure drop is negligible small at low rates of entrainment; calculation: see section 6.1.3). If the pressure drop of the moving bed is assumed to be equal to the one at minimum fluidization, the height difference of the moving bed $\Delta H_{SP,bed}$ can be calculated using the Ergun [8] equation 2.1:

$$\frac{\Delta P_{H_{SR}-H_{FB}}}{\Delta H_{SP,bed}} = 150 \frac{\eta_f}{d_p^2} \frac{\left(1-\epsilon_{mf}\right)^2}{\epsilon_{mf}^3} U_{mf} + 1.75 \frac{\rho_f}{d_p} \frac{\left(1-\epsilon_{mf}\right)}{\epsilon_{mf}^3} U_{mf}^2 \tag{6.5}$$

Here, η_f is the gas viscosity, d_p the mean particle size, ϵ_{mf} the bed porosity at minimum fluidization $(= 1 - c_{V,mf})$ and ρ_f the density of the gas. The minimum fluidization velocity U_{mf} is estimated by a literature approach given in equation 2.2. As equation 6.5 shows, knowledge about the pressure drop in the fluidized bed is necessary to estimate the amount of solids standing in the standpipe.

In addition to the solids standing, there is also an amount of solids continuously falling in the standpipe. It is assumed that these particles fall down at their terminal velocity. The sum of the masses of standing and falling solids is the overall mass of solids in the standpipe, which can be calculated according to

$$m_{SP} = A_{SP}\rho_b\Delta H_{SP,bed} + \frac{G_s A_0}{U_t}\left(H_{FB} - H_{LS} - \Delta H_{SP,bed}\right).\qquad(6.6)$$

The mass of the solids standing is hereby calculated from the height of the moving bed $\Delta H_{SP,bed}$, the cross-sectional area of the standpipe A_{SP} and the bulk density ρ_b. The amount of solids material falling down results from the solids circulation rate G_s, the cross-sectional area of the fluidized bed A_0, the terminal velocity of the particles U_t and the height difference from loop-seal to the top of the standpipe minus the height of the moving bed $H_{SP} - H_{LS} - \Delta H_{SP,bed}$.

If the mass of solids in the cyclone and other parts is assumed to be negligible small, the overall mass of bed material in the system m_{total} results as follows:

$$m_{total} = m_{FB} + m_{SP}\qquad(6.7)$$

The mass balances are derived on the assumption of a fluidized bed system containing the fluidized bed, a cyclone, a standpipe and a loop-seal only. If the whole system is more complex, the mass balances must be adapted accordingly.

6.1.3 Pressure Drop

The freeboard above the transport disengaging height of a turbulent fluidized bed is defined by a dilute up-flow of particles. Thus, the flow behavior in this part is similar to pneumatic conveying. In the dense zone of the fluidized bed larger solids concentrations occur. The pressure drop in this part of a fluidized bed is mainly affected by the hydrostatic forces induced by the solids. For this reason, equation 2.5 can be used to give a rough estimation of the fluidized bed pressure drop. In addition to the hydrostatic pressure drop, acceleration effects of the particles have been found to largely impact the pressure drop in the bottom section of the fluidized bed [110, 205].

In pneumatic conveying the pressure drop is the sum of the hydrostatic pressure drops, friction induced pressure drops and acceleration pressure drops of the solids and the fluid, respectively (based on [209–211]):

$$\Delta P = \Delta P_{hyd,f} + \Delta P_{hyd,s} + \Delta P_{fric,f} + \Delta P_{fric,s} + \Delta P_{acc,f} + \Delta P_{acc,s}\qquad(6.8)$$

In this work it is assumed that the same pressure drops as they occur in pneumatic conveying are responsible for the pressure loss in fluidized beds. Thereby, the fluid terms are small in comparison to the pressure drops of solids hydrostatics and acceleration. The different terms of equation 6.8 for a fluidized bed are defined in the following.

6.1.3.1 Hydrostatic pressure drop

The hydrostatic pressure drops of the fluid

$$\Delta P_{hyd,f} = \rho_f \epsilon g \Delta h \tag{6.9}$$

and the solids

$$\Delta P_{hyd,s} = \rho_p \left(1 - \epsilon \right) g \Delta h \tag{6.10}$$

depend on their densities ρ_f and ρ_p, the bed porosity ϵ, the gravitational acceleration g and the height difference Δh.

6.1.3.2 Pressure drop due to friction

The definitions of the pressure drops induced by friction of fluid and solids partially differ from each other in literature. According to Monazam *et al.* [212] the friction terms of the pressure drop in a circulating fluidized bed can be defined as derived by Hinkle [211] for pneumatic conveying. In contrast to this, Klinzing [209] defines the frictional pressure drops with a factor of 0.5 for the fluid and 0.25 for the solids in addition to the equations given by Monazam *et al.* [212]. In combination with the correlations proposed for the gas friction factor of each source, the approach of Sommerfeld *et al.* [210] is similar to the one of Monazam *et al.* [212]. Additionally, for high solids concentrations Sommerfeld *et al.* [210] consider the bed porosity in their equation.

In this work the pressure drop of gas friction is defined as

$$\Delta P_{fric,f} = 2 f_g \frac{\Delta h}{D} \rho_f \epsilon U_f^{\,2}, \tag{6.11}$$

which follows the approach of Monazam *et al.* [212] under consideration of the bed porosity ϵ as proposed by Sommerfeld *et al.* [210]. The correlation used for the gas friction factor is given by Monazam *et al.* [212] with

$$f_g = 0.0014 + 0.125 Re_D^{\,-0.32}. \tag{6.12}$$

The Reynolds number Re_D is related to gas properties, the gas velocity and the fluidized bed diameter D. The gas velocity U_f is assumed to equal the superficial gas velocity U_0.

The pressure drop induced by solids friction is defined as given by Monazam *et al.* [212] with

$$\Delta P_{fric,s} = 2f_s\frac{\Delta h}{D}\rho_p\left(1 - \epsilon\right)U_p^{2}. \tag{6.13}$$

The friction factor is defined according to Sommerfeld *et al.* [210] as:

$$f_s = 0.0285\frac{(gD)^{0.5}}{U_p} \tag{6.14}$$

Because the real velocity of the particles U_p varies over the beds cross-section it is assumed to equal the velocity of the bubbles U_B in the dense zone and to be the difference of superficial gas velocity and terminal velocity of the particles $U_0 - U_t$ in the freeboard zone:

$$U_p = \begin{cases} U_B, & h \leq H_{exp} \\ U_0 - U_t, & h > H_{exp} \end{cases} \tag{6.15}$$

The velocity of the bubbles can be calculated combining equations 5.13 and 5.14.

6.1.3.3 Pressure drop due to acceleration

The acceleration pressure drop of the fluid is assumed to be negligible:

$$\Delta P_{acc,f} = 0 \tag{6.16}$$

Nevertheless, in the dense zone of the fluidized bed the bubble coalescence and flow phenomena might have influence on fluid velocities resulting in acceleration and deceleration of the fluid.

Following Hinkle [211] and Sommerfeld *et al.* [210] the acceleration pressure drop in pneumatic conveying can be calculated similar to

$$\Delta P_{acc,s} = \rho_p\left(1 - \epsilon\right)\Delta U_p^{2}. \tag{6.17}$$

As mentioned before, in a fluidized bed real particle velocities vary over the bed cross-section. Furthermore, in a turbulent fluidized bed, large amounts of particles are accelerated in the dense zone of the fluidized bed. These particles rise with the rise of bubbles and are ejected into the transition to the freeboard zone, where they partially decelerate. The decelerated and not entrained particles move downwards in the fluidized bed. The down stream of particles mostly occurs close to the fluidized bed wall. A continuous characteristic fluidized bed flow pattern is the result. At steady state the net mass flow over the cross-section of a fluidized bed should equal the mass flow of entrained particles above the solids return and should be zero below. Thus, one could assume that the acceleration pressure drop should only be caused by the entrained particles that are returned to the fluidized bed. Nevertheless, the pressure drop in the dense zone is larger and at transition to the freeboard zone lower than expected, as measurements show. These findings might result from a greater influence of the beds internal acceleration and deceleration effects. For determination of the pressure drop induced by acceleration and deceleration knowledge about the amount of particles rising and their velocity is necessary. These information cannot be yielded from the measurement techniques used in this work. For this reason, acceleration and deceleration are neglected in the model predictions of the pressure drop. This should not influence results of the total pressure drop over the whole bed because particles accelerated at the fluidized bed bottom are decelerated at higher heights. Therefore, the acceleration pressure drop equals to the deceleration pressure drop.

6.1.4 Calculation Flowchart

The calculation procedure of the model is shown in a flowchart in figure 6.1. It shows the different input parameters that are necessary to predict the axial solids concentration and the pressure drop over the whole facility. The input parameters can be divided into bed material properties, fluidized bed plant properties and operational properties.

Subsequently, characteristic fluidized bed properties are calculated like the minimum fluidization velocity, the rate of entrainment, mean solids concentration in the dense and the freeboard zone, and bubble properties. The bed expansion height is determined iterative by calculation of the axial solids concentration profile, the fluidized bed pressure drop and by solving the mass balance equations. If the total mass in the system equals the desired bed mass, the final concentration and pressure profiles are received.

The model is proposed for a fluidized bed facility consisting of the fluidized bed, a cyclone, a standpipe and a loop-seal only. For more complex systems the model must be adjusted accordingly.

6.2 Comparison to Experimental Data

For validation of the proposed model, cases at different fluidization conditions predicted by the model are compared to solids concentrations measured by capacitance probe and the collected pressure data.

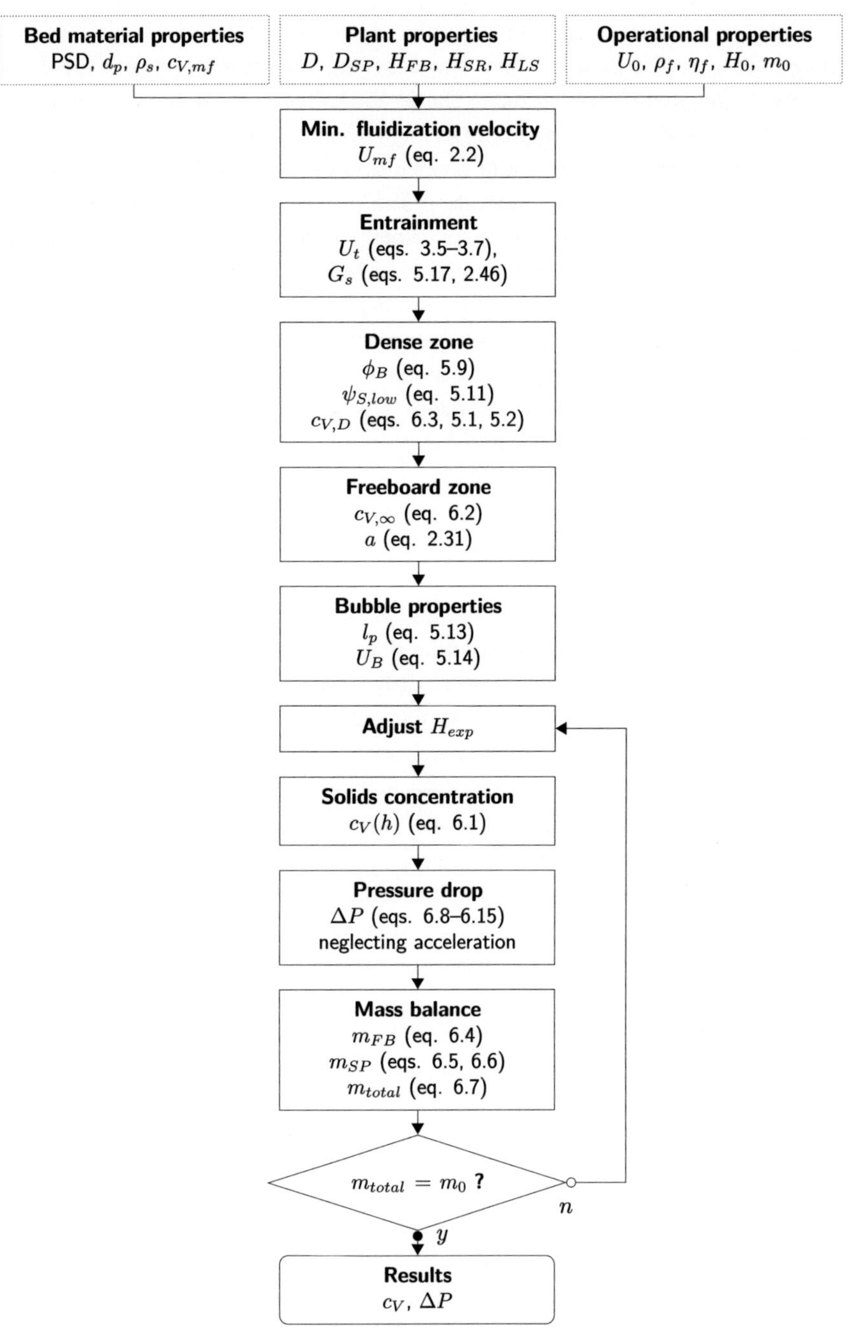

Fig. 6.1: Calculation flowchart of the proposed model.

6.2.1 Solids Concentration

In figure 6.2, the cross-sectionally and temporally averaged solids concentrations measured are plotted for two different superficial gas velocities in the dense zone of the fluidized bed. The predictions of the solids concentrations by the model are in a good agreement to the measurement data. With increasing superficial gas velocity lower solids concentrations are measured and predicted. This results from an increasing bubble hold-up and more gas streaming through the suspension phase. The expansion height of the bed slightly increases. Furthermore, more bed material can be found at larger heights due to the eruption of larger and faster bubbles at the bed surface. The two superficial gas velocities shown in the figure belong to the bubbling fluidized bed regime. As the results show, the model is capable of predicting fluidization behavior in the bubbling state.

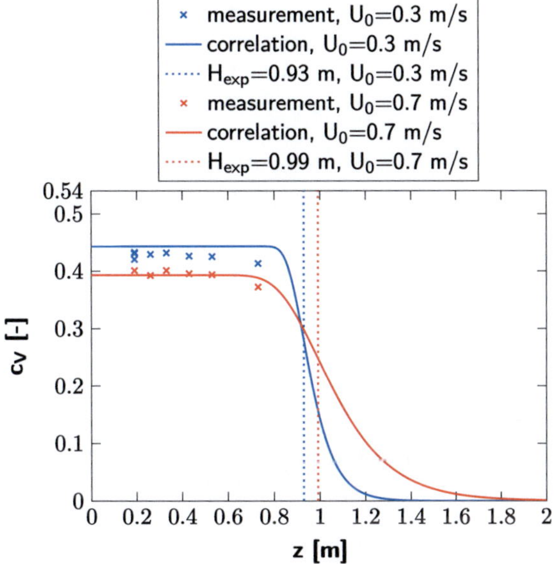

Fig. 6.2: Solids concentrations measured by capacitance probe and predicted for sand B1 in dependence of the height above the gas distributor at $D = 0.4\,\text{m}$ and $H_0 = 0.8\,\text{m}$.

At the same superficial gas velocities at a larger bed diameter the model gives a slight overestimation of the solids concentrations in the dense zone as it can be seen in figure 6.3. The solids concentrations predicted at a height of 1.23 m agree well with the measurements. In the model, solids concentrations are calculated from the bubble phase hold-up and the solids concentration of the dense phase according to the two-phase theory. The latter is calculated under the assumption of the existence of a low and a high concentration dense phase. These assumptions can lead to deviations of the predicted values from the measured ones. At a larger superficial gas velocity of $1.25\,\mathrm{m\,s^{-1}}$ around the transition to turbulent fluidization deviations from the measurements can still be found. Nevertheless, the solids concentrations are predicted at a high sufficiency for larger bed heights. At a low static bed height and large superficial gas velocities a dense zone of the fluidized bed could not be distinguished from the transition into the freeboard zone anymore, as measurements in the CFB400 at an aspect ratio of 1 showed.

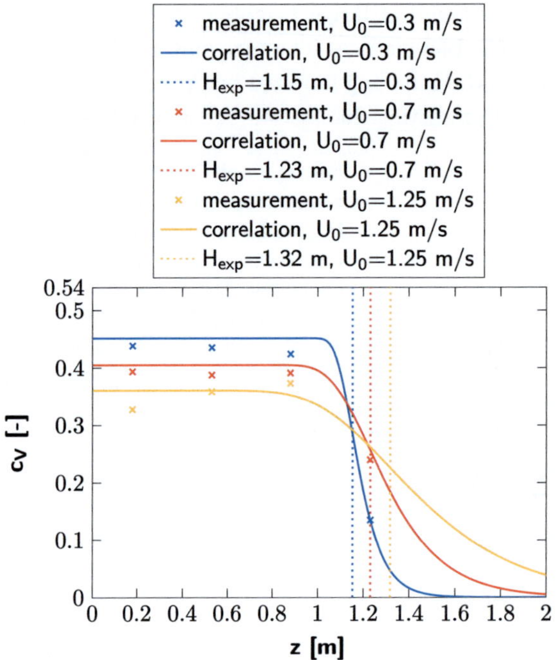

Fig. 6.3: Solids concentrations measured by capacitance probe and predicted for sand B1 in dependence of the height above the gas distributor at $D = 1\,\mathrm{m}$ and $H_0 = 1\,\mathrm{m}$.

6.2.2 Pressure Profiles

The modeling approach is built on solids concentration measurements carried out with capacitance probes. For this reason, only small deviations of the model predictions from the measured data is expected and can be found in the figures 6.2 and 6.3. As a further validation parameter pressure data are used of which the model is completely independent. The absolute pressures in the fluidized bed are calculated from the differential pressures under the assumption of ambient conditions at the fluidized bed exit for both cases, measurement and prediction, respectively.

As mentioned before, acceleration effects in the dense zone of the fluidized bed have been neglected due to unknown particle velocities. Nevertheless, these effects are significantly influencing the pressure drop as reported in literature [110, 205]. The measured and predicted absolute pressures in the CFB400 for an aspect ratio around one are shown in figure 6.4 for four different superficial gas velocities with (a) being at bubbling fluidization, (b) at turbulent fluidization and (c) and (d) at fast fluidization. The predicted pressure curves in (a) and (b) slightly overestimate the pressures in the freeboard region of the bed. A steeper slope can be found for the predicted curves, especially in figure (b). The deviation of the predictions from the measurement means that fewer bed material is actual entrained than predicted by the model or the pressure drop resulting from friction is overestimated. The latter can be excluded due to the friction related pressure drop being comparably small (pressure drop due to friction is around 6 % of total bed pressure drop). For the static bed height investigated in figure 6.4 lower entrainment rates have been found than for larger static bed heights. To determine entrainment rates equation 5.17 was used in the model. This equation was fitted to particle entrainment data yielded from measurements of static bed heights above 0.8 m where particle entrainment becomes independent of it. For this reason, the applicability of the proposed model is limited to these bed heights. The result is an overestimation of the entrainment and consequently of the pressure drop in the freeboard zone as shown in figure 6.4.

According to the mass balance, if more bed material is predicted in the freeboard zone of the fluidized bed, fewer material is the result in the dense zone. Thus, it can be expected that the bed expansion height is underestimated in this case. Nevertheless, the predicted pressure drop over the whole fluidized bed equals the one measured in figure 6.4 (a)–(c). This means, that the determination of the total bed pressure drop under consideration of the hydrostatic pressure drops and the frictional pressure drops can be assumed to be sufficient. Deviations in the curves can be related to the larger entrainment rates predicted and the acceleration and deceleration of particles in the bed.

To give a rough estimation of the influence of acceleration on pressure drop in the dense zone the acceleration pressure drop between two points was approached as follows:

$$\Delta P_{acc,12} = 0.5\kappa \left(U_{p,2}{}^2 - U_{p,1}{}^2 \right) \rho_s \left(c_{V,D} - c_{V,\infty} \right) \tag{6.18}$$

The particle velocities have been assumed to equal the velocities of the bubbles at the two points. This equation is derived from the Bernoulli equation under

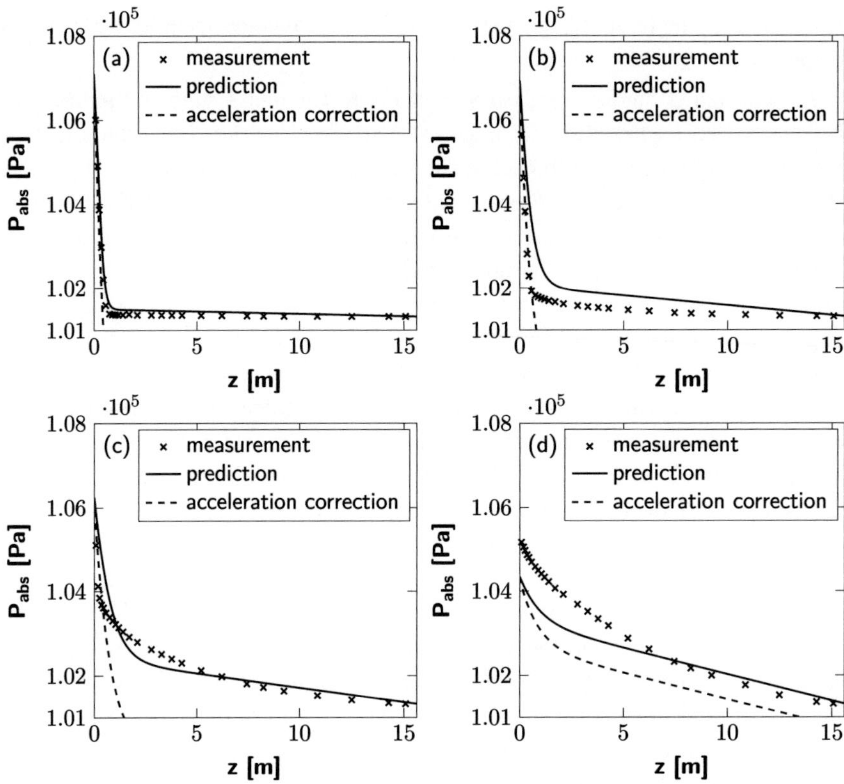

Fig. 6.4: Predicted and measured pressure profiles for sand B1 at $D = 0.4\,\mathrm{m}$ and $H_0 = 0.39\,\mathrm{m}$ with (a) $U_0 = 0.52\,\mathrm{m\,s^{-1}}$, (b) $U_0 = 1.52\,\mathrm{m\,s^{-1}}$, (c) $U_0 = 2.52\,\mathrm{m\,s^{-1}}$, and (d) $U_0 = 3.57\,\mathrm{m\,s^{-1}}$.

the assumption of the acceleration pressure drop being the difference of the dynamic pressures between two points in the bed. Because a fluidized bed consists of regions of up-flowing and down-flowing solids, it must be considered that only a part of the bed material rises and causes a change of the dynamic pressure. For this reason, the parameter κ was added to equation 6.18. Under the assumption that the fluidized bed is in complete circulation in the turbulent regime (solids rising in the center and descending in the wall region) around 50 % of the particles rise and a value of 0.5 can be assumed for κ. This might not be the case in bubbling fluidization at low superficial gas velocities where only few bubbles rise and carry less solids. Thus, in this flow regime an overestimation of the acceleration pressure drop using equation 6.18 can be the result.

The corrections of the pressure curves are shown as dashed lines in the plots of figure 6.4. Because the deceleration of the particles is not considered in these curves, the total bed pressure drop predicted by these curves is larger than the one measured. The validity of these curves is only given in the dense zone of the fluidized bed. As figure 6.4 (b) and (c) show, the correction of the absolute pressures by the acceleration pressure drop calculated with equation 6.18 gives a very accurate prediction of the pressure drop in the dense zone of the fluidized bed, where the pressure gradient is large due to high solids concentrations. In (a) the acceleration correction gives a slight underestimation of the absolute pressure. As explained before, this can be related to the rise of bubbles and the resulting circulation pattern in the fluidized bed for which a value of 0.5 is too large for κ.

In figure 6.4 (d) in the clear fast fluidization regime the predicted pressure drops do not agree to the measurements. The predicted total pressure drop is lower than the one measured which might be related to the overestimation of the entrainment rate. The model predicts more material to stand in the standpipe resulting in lower amounts in the fluidized bed.

Figure 6.5 shows the predicted and measured pressure curves at the same conditions as in figure 6.4 but with a larger static bed height. In this case the correlation of solids entrainment rates gives sufficient predictions. From bubbling fluidization up to fast fluidization the model gives absolute pressures being in a good agreement to the measurements. The deviations related to acceleration and deceleration effects increase with increasing superficial gas velocity. In the dense zone of the fluidized bed the model gives an underestimation of the pressure drop. In contrast to this, in the region of transition between dense and freeboard zone the pressure drop of the model is overestimated especially in the state of fast fluidization. This can be related to the deceleration of particles in this region.

The predicted total bed pressure drops agree very well with the measurements. Thus, it can be assumed that the predictions of the entrainment and of the bed material holdup in the standpipe are valid.

In the bubbling regime in figure 6.5 (a) the fluidized bed pressure is represented well by the model. The acceleration correction of the pressure gives a slight overestimation of the pressure drop, as it is the case in figure 6.4 (a) as well. In the plots (b)–(d) of figure 6.5 the acceleration correction gives a sufficient description of the pressure drop in the dense zone and thus the absolute pressure there.

As it can be seen in the plots, above the region of deceleration (transition from dense zone to freeboard zone) the predicted pressure curves merge with the measurements. In this region of the freeboard a continuous flow of particles is entrained.

In a bed of a larger diameter similar prediction behavior can be found. Figure

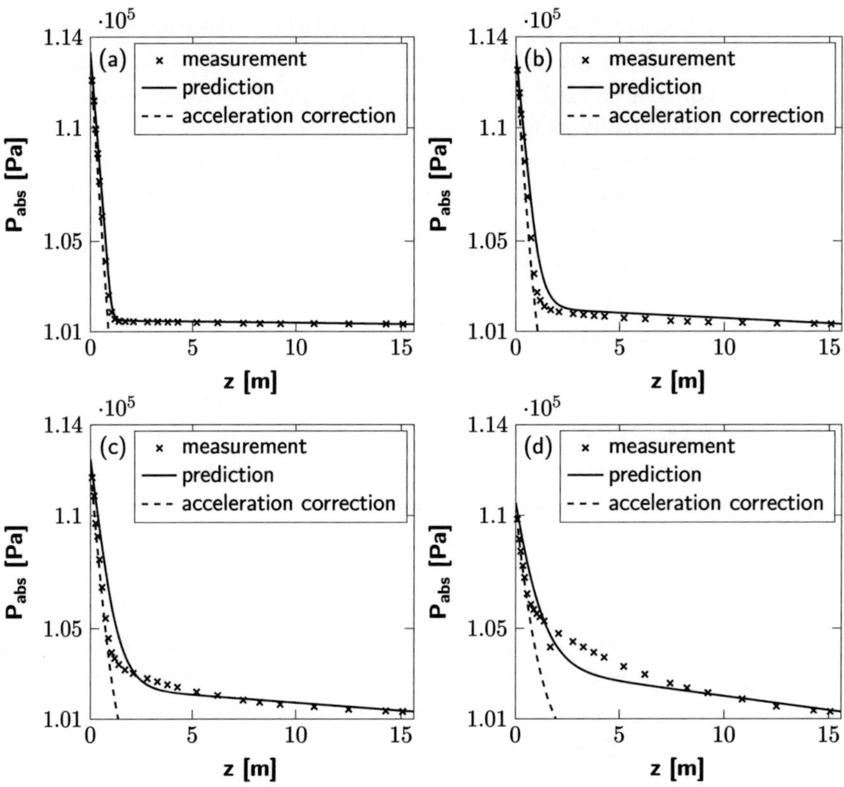

Fig. 6.5: Predicted and measured pressure profiles for sand B1 at $D = 0.4\,\text{m}$ and $H_0 = 0.82\,\text{m}$ with (a) $U_0 = 0.47\,\text{m s}^{-1}$, (b) $U_0 = 1.53\,\text{m s}^{-1}$, (c) $U_0 = 2.54\,\text{m s}^{-1}$, and (d) $U_0 = 3.55\,\text{m s}^{-1}$.

6.6 shows the predicted pressure curves in comparison to the pressure measurements in the bubbling regime in (a)-(c) up to the regime transition to the turbulent regime in (d). At a low superficial gas velocity in (a) the measurements are represented by the model with high accuracy. Thus, it can be assumed from the pressure data prediction that solids concentration profiles must agree as well, which was proven in figure 6.3. In figure 6.6 (b)–(d) a deviation of the predicted curves from the pressure measurements can be seen. The deviation increases at larger superficial gas velocities due to the increasing influence of particle acceleration and deceleration effects. Nevertheless, the predicted total bed pressure drop agrees with the measured one.

The curves of the acceleration correction give an overestimation of the pressure drop in all graphs in figure 6.6. Thus, the predicted absolute pressure is lower

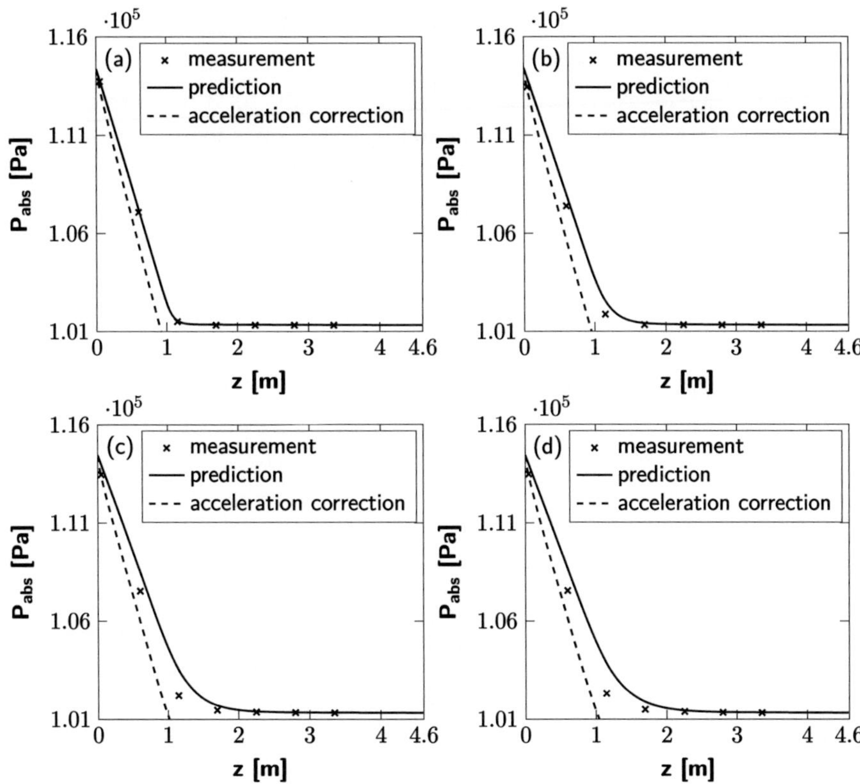

Fig. 6.6: Predicted and measured pressure profiles for sand B1 at $D = 1\,\mathrm{m}$ and $H_0 = 1\,\mathrm{m}$ with (a) $U_0 = 0.3\,\mathrm{m\,s^{-1}}$, (b) $U_0 = 0.7\,\mathrm{m\,s^{-1}}$, (c) $U_0 = 1.1\,\mathrm{m\,s^{-1}}$, and (d) $U_0 = 1.25\,\mathrm{m\,s^{-1}}$.

than measured. This can be related to the value of κ in equation 6.18, which is chosen too large for the bubbling state. The fluidization pattern in the FB1000 is affected by the rise of bubbles. In the turbulent regime the bubbles can be found at the whole cross-section with lower probability at the fluidized bed wall. In contrast to this, in bubbling fluidization the bubbles follow flow paths [55]. Thus, regions of deprived bubble flow occur where particles are less influenced by the rise of the bubbles. In this case, it cannot be assumed that 50 % of the bed material in a cross-section are influenced by the bubble flow and rise with the velocity of the bubbles. The result is a lower value of κ, which increases with the superficial gas velocity. For this reason, the influence of the acceleration pressure drop increases slightly. The increase of κ, which is directly related to the bubble behavior, agrees with findings of the axial solids dispersion coefficient. The axial solids dispersion coefficient increases in the bubbling regime as well, indicating larger solids movements [140].

It can be summarized that the model predicts the total pressure drop of a fluidized bed with high accuracy. Due to the increasing influence of acceleration and deceleration effects, the deviation of the predicted pressure profile from the measurements increases with increasing superficial gas velocity. Nevertheless, in the dense zone of a turbulent fluidized bed the usage of equation 6.18 gives a sufficient approach for the prediction of the acceleration pressure drop.

6.3 Model Coupling

The proposed model in this work is a fluid dynamic model capable of predicting the solids hold-up of a turbulent fluidized bed. For modeling of chemical reaction behavior in a fluidized bed the fluid dynamic sub-model must be coupled with a kinetic sub-model. This can be done by calculation of the mass balances of each phase and each component.

The fuel reactor in the chemical looping combustion process can be operated as turbulent fluidized bed. The model was developed for application in this process. In the fuel reactor the oxygen carrier - which circulates between fuel and air reactor - is depleted. The oxygen released by the oxygen carrier reacts with a fuel, which is fed to the reactor. In the proposed two-phase model the bubble phase is considered to be empty of solid material. For this reason, it is assumed that no reaction takes place in the bubble phase. For the mass balance of the gas component i in the bubble phase gb

$$\frac{\partial C_{i,gb}}{\partial t} = -U_{gb}\frac{\partial C_{i,gb}}{\partial z} - k_{bs}a_b\left(C_{i,gb} - C_{i,gs}\right) \tag{6.19}$$

results for the non-steady state. It describes the temporal change of the concentration C of component i in dependence of axial concentration variations and the inter-phase mass transfer of the gas between bubble (gb) and suspension phase (gs). Hereby, the velocity U_{gb} is the velocity of the gas that moves in the bubble phase. As summarized in table 2.6 different approaches for the mass transfer coefficient between both phases are available.

In contrast to the bubble phase, the mass balance of the gas component i in the suspension phase gs is additionally influenced by the axial gas dispersion and the chemical reaction behavior:

$$\frac{\partial C_{i,gs}}{\partial t} = D_{g,z}\frac{\partial^2 C_{i,gs}}{\partial z^2} - U_{gs}\frac{\partial C_{i,gs}}{\partial z} + k_{bs}a_b\left(C_{i,gb} - C_{i,gs}\right) + R_{ij} \tag{6.20}$$

The axial gas dispersion coefficient $D_{g,z}$ is given by different correlations in literature as listed in table 2.3. The velocity of the gas in the suspension phase U_{gs} is the interstitial gas velocity. It can be calculated using the approach in equation 5.16. The chemical reaction behavior in equation 6.20 is given by the term R_{ij}.

Because the bubble phase is assumed to be empty of solids, mass transfer of solids into the bubble phase is not considered. The mass balance of the solid component i in the suspension phase ss can be described as follows:

$$\frac{\partial C_{i,ss}}{\partial t} = D_{s,z}\frac{\partial^2 C_{i,ss}}{\partial z^2} - U_{ss}\frac{\partial C_{i,ss}}{\partial z} + R_{ij} \tag{6.21}$$

Here, the mass balance is defined by the axial solids dispersion, axial concentration variations and the chemical reaction term. Lee and Kim [140] introduced an approach for the determination of the axial solids dispersion coefficient for particles of Geldart's group B. Because oxygen carriers in chemical looping combustion belong to this group of particles, their correlation can be used. At steady state the velocity of solids U_{ss} is linked to the mass flow of solid material circulating between the reactors. Locally, larger mass flows can be found in the fluidized bed but due to steady state they compensate each other over the cross-section of the fluidized bed to the circulating mass flow.

The final modeling approach is not only limited to the chemical looping process. It can also be used for prediction of conversion rates in other fluidized bed processes using particles of Geldart's group B. If the process requires a specific facility set-up, the fluid dynamic sub-model must be adjusted accordingly.

7 Conclusions

Detailed information about fluidization behavior of turbulent fluidized beds using particles of Geldart's group B are rare. Only few studies concentrate on this fluidization regime with this kind of particles having partially contradictory statements. Nevertheless, knowledge about the fluidization behavior is urgently needed for plant design and scale-up of uprising carbon capture and storage technologies like the chemical looping combustion to contribute to mitigation of climate change. For this reason, an in-depth experimental study was carried out for the development of a fluid dynamic model which predicts fluidization behavior in turbulent fluidized beds with particles belonging to Geldart's group B.

The experiments have been conducted in fluidized beds of different sizes ranging from 0.05 m to 1 m in diameter. Thus, it was possible to capture wall effects influencing the fluidization behavior, which is of high interest for fluidized bed scale-up. The superficial gas velocities have been adjusted between velocities close to minimum fluidization velocity up to $5\,\mathrm{m\,s^{-1}}$, which is clearly beyond the turbulent fluidized bed regime at fast fluidization. This allowed conclusions about the similarities and differences of the fluidization regimes. Different bed materials belonging to Geldart's group B have been characterized and used for investigations in the fluidized bed facilities.

Pressure fluctuations have been evaluated to identify the fluidized bed regimes according to their definitions in literature. The transition from bubbling to turbulent fluidization was found to be dependent on bed material properties (particle size distribution, particle density), fluidized bed properties (bed diameter, static bed height), the measurement method (measurement location, absolute or differential pressure measurement) and the gas properties (gas density, gas viscosity). The dependencies have been found to agree with literature findings. Because correlations in literature for the prediction of the transition velocity only consider few of the parameters mentioned above, a new correlation was introduced in this work. For the transition from turbulent to fast fluidization similar trends can be assumed.

Frequency analysis of the pressure signals showed that the behavior of the intensities of the power spectra equals the behavior of the standard deviation of the pressure signals. Thus, frequency analysis can be used as alternative method for the determination of the transition velocity from bubbling to turbulent fluidization. The frequency analysis method gives slightly lower values for the transition velocities than the standard deviation method.

For the determination of local solids concentrations and bubble properties capacitance probes have been used. The probability density distributions of the local solids concentrations showed a clear two-phase behavior even in the turbulent fluidized bed regime. Moreover, the suspension phase was found to form a low and a high concentration suspension phase. Typical radial solids concentration profiles could be found at bubbling fluidization with bubbles developing preferred flow paths. Regions of concentrated bubble flow have been developed near the fluidized bed wall at the gas distributor. They move towards the bed center with increasing height in the fluidized bed until they merge. In contrast to this, in the turbulent fluidized bed parabolic concentration profiles can be found at each height.

By evaluation of the probability density distributions of the solids concentrations a phase border between bubble and suspension phase was defined. Using this phase border the bubble phase hold-ups were determined. They showed similar trends in their radial profiles as the profiles of the solids concentration proving the flow in preferred paths in the bubbling regime. The cross-sectionally averaged bubble phase hold-ups measured in three plants have been correlated. A strong dependence on the superficial gas velocity and the bed diameter have been found. Due to the rising amount of bubbles with increasing superficial gas velocity, the bubble hold-up increases. Furthermore, the wall effects in small fluidized beds lead to a larger bubble phase hold-up in comparison to beds having larger diameters.

Vertical bubble sizes (pierced lengths) and bubble velocities have been evaluated from the capacitance probe measurements. Both could be described by logarithmic normal distributions, which broadened at increasing superficial gas velocities. Bubbles having larger pierced lengths, rising at larger velocities occurred at larger superficial gas velocities. This happened even beyond the point of transition into the turbulent regime. Thus, in contrast to the behavior of particles of Geldart's group A, in beds of group B particles, bubble growth and coalescence continuous further and bubble break-up does not overtake growth processes. The averaged pierced lengths have been correlated in dependence of superficial gas velocity, height above the distributor and bed diameter. In addition to this, the velocity of bubbles has been found to converge to a limiting maximum bubble velocity in dependence on bubble size. A further correlation for the bubble velocity in dependence on the pierced length was introduced. Calculation of the drag coefficient of the bubbles showed, that they must change shape by transition into the turbulent regime from the spherical cap bubble in the bubbling regime to a vertical stretched shape. This results in the measured decrease of the mean pressure fluctuation amplitude by entering the turbulent regime.

Finally, a model was introduced for the prediction of the fluidization behavior of turbulent fluidized beds with particles of Geldart's group B. The model is one dimensional and based on the two-phase approach. The bubble phase is assumed to be empty of solids. Gas streams through the bubble phase as well as through the suspension phase. The modeling approach is based on the correlations derived in this work. It is capable of predicting parameters like the axial solids concentration distribution, pressure drops, entrainment rates, bubble properties and phase hold-ups. The solids concentrations predicted by the model agree well with measured concentrations. Due to acceleration and deceleration effects in the fluidized bed, it was not possible to predict the pressure curves with high accuracy over the entire measurement range. Nevertheless, by introduction of an acceleration correction in the dense zone of the bed the pressure drop could be predicted with a good accuracy.

To summarize, this work delivers detailed information about fluidization regime transition and identification and understanding of phase and bubble behavior in bubbling and turbulent fluidized beds. The model is capable of predicting the important fluidization properties and can be used for plant design and scale-up. This work extends the current knowledge of turbulent fluidized bed fluidization behavior and contributes to the fundamental understanding of fluidized bed technology.

Bibliography

[1] VijayaVenkataRaman, S., Iniyan, S., and Goic, R. "A review of climate change, mitigation and adaptation". In: *Renewable and Sustainable Energy Reviews* 16.1 (2012), pp. 878–897.

[2] Meinshausen, M., Smith, S. J., Calvin, K., Daniel, J. S., Kainuma, M. L. T., Lamarque, J.-F., Matsumoto, K., Montzka, S. A., Raper, S. C. B., Riahi, K., Thomson, A., Velders, G. J. M., and van Vuuren, D. P. "The RCP greenhouse gas concentrations and their extensions from 1765 to 2300". In: *Climatic Change* 109.1-2 (2011), pp. 213–241.

[3] Jackson, R. B., Le Quéré, C., Andrew, R. M., Canadell, J. G., Korsbakken, J. I., Liu, Z., Peters, G. P., and Zheng, B. "Global energy growth is outpacing decarbonization". In: *Environmental Research Letters* 13.12 (2018), p. 120401.

[4] Olivier, J. and Peters, J. *Trends in Global CO_2 and Total Greenhouse Gas Emissions: Report 2019*. Tech. rep. 4068. The Hague: PBL Netherlands Environmental Assessment Agency, 2020.

[5] Adánez, J., Abad, A., Mendiara, T., Gayán, P., de Diego, L. F., and García-Labiano, F. "Chemical looping combustion of solid fuels". In: *Progress in Energy and Combustion Science* 65 (2018), pp. 6–66.

[6] Bi, H., Ellis, N., Abba, I., and Grace, J. "A state-of-the-art review of gas–solid turbulent fluidization". In: *Chemical Engineering Science* 55.21 (2000), pp. 4789–4825.

[7] Du, B., Fan, L. S., Wei, F., and Warsito, W. "Gas and solids mixing in a turbulent fluidized bed". In: *AIChE Journal* 48.9 (2002), pp. 1896–1909.

[8] Ergun, S. "Fluid flow through packed columns". In: *Chemical Engineering Progress* 48.2 (1952), pp. 89–94.

[9] Yang, W.-C. "Bubbling Fluidized Beds". In: *Handbook of Fluidization and Fluid-Particle Systems*. Ed. by Yang, W.-C. 1st ed. New York, Basel: Marcel Dekker, Inc., 2003, pp. 53–112.

[10] Wen, C. Y. and Yu, Y. H. "A generalized method for predicting the minimum fluidization velocity". In: *AIChE Journal* 12.3 (1966), pp. 610–612.

[11] Davidson, J. F., Harrison, D., and Carvalho, J. R. F. G. D. "On the Liquidlike Behavior of Fluidized Beds". In: *Annual Review of Fluid Mechanics* 9 (1977), pp. 55–86.

[12] Basu, P. *Circulating Fluidized Bed Boilers*. Cham, Heidelberg, New York, Dordrecht´, London: Springer International Publishing, 2015.

[13] Karimipour, S. and Pugsley, T. "A critical evaluation of literature correlations for predicting bubble size and velocity in gas-solid fluidized beds". In: *Powder Technology* 205.1-3 (2011), pp. 1–14.

[14] Baeyens, J. and Geldart, D. "An investigation into slugging fluidized beds". In: *Chemical Engineering Science* 29.1 (1974), pp. 255–265.

[15] Geldart, D. "Types of gas fluidization". In: *Powder Technology* 7.5 (1973), pp. 285–292.

[16] Abad, A., Adánez, J., Gayán, P., de Diego, L. F., García-Labiano, F., and Sprachmann, G. "Conceptual design of a 100 MW_{th} CLC unit for solid fuel combustion". In: *Applied Energy* 157 (2015), pp. 462–474.

[17] Abad, A., Adánez, J., García-Labiano, F., de Diego, L. F., Gayán, P., and Celaya, J. "Mapping of the range of operational conditions for Cu-, Fe-, and Ni-based oxygen carriers in chemical-looping combustion". In: *Chemical Engineering Science* 62.1-2 (2007), pp. 533–549.

[18] Abad, A., Adánez-Rubio, I., Gayán, P., García-Labiano, F., de Diego, L. F., and Adánez, J. "Demonstration of chemical-looping with oxygen uncoupling (CLOU) process in a 1.5 kW_{th} continuously operating unit using a Cu-based oxygen-carrier". In: *International Journal of Greenhouse Gas Control* 6 (2012), pp. 189–200.

[19] Adánez, J., Gayán, P., Celaya, J., de Diego, L. F., García-Labiano, F., and Abad, A. "Chemical Looping Combustion in a 10 kW_{th} Prototype Using a CuO/Al_2O_3 Oxygen Carrier: Effect of Operating Conditions on Methane Combustion". In: *Industrial & Engineering Chemistry Research* 45.17 (2006), pp. 6075–6080.

[20] Adánez-Rubio, I., Pérez-Astray, A., Mendiara, T., Izquierdo, M. T., Abad, A., Gayán, P., de Diego, L. F., García-Labiano, F., and Adánez, J. "Chemical looping combustion of biomass: CLOU experiments with a Cu-Mn mixed oxide". In: *Fuel Processing Technology* 172 (2018), pp. 179–186.

[21] Cuadrat, A., Abad, A., García-Labiano, F., Gayán, P., de Diego, L. F., and Adánez, J. "Effect of operating conditions in Chemical-Looping Combustion of coal in a 500 W_{th} unit". In: *International Journal of Greenhouse Gas Control* 6 (2012), pp. 153–163.

[22] Gayán, P., Adánez-Rubio, I., Abad, A., de Diego, L. F., García-Labiano, F., and Adánez, J. "Development of Cu-based oxygen carriers for Chemical-Looping with Oxygen Uncoupling (CLOU) process". In: *Fuel* 96 (2012), pp. 226–238.

[23] Gayán, P., Abad, A., de Diego, L. F., García-Labiano, F., and Adánez, J. "Assessment of technological solutions for improving chemical looping combustion of solid fuels with CO_2 capture". In: *Chemical Engineering Journal* 233 (2013), pp. 56–69.

[24] Mendiara, T., de Diego, L. F., García-Labiano, F., Gayán, P., Abad, A., and Adánez, J. "On the use of a highly reactive iron ore in Chemical Looping Combustion of different coals". In: *Fuel* 126 (2014), pp. 239–249.

[25] Mendiara, T., Izquierdo, M. T., Abad, A., Gayán, P., García-Labiano, F., de Diego, L. F., and Adánez, J. "Mercury release and speciation in chemical looping combustion of coal". In: *Energy and Fuels* 28.4 (2014), pp. 2786–2794.

[26] Pérez-Vega, R., Abad, A., García-Labiano, F., Gayán, P., de Diego, L. F., and Adánez, J. "Coal combustion in a 50 kW_{th} Chemical Looping Combustion unit: Seeking operating conditions to maximize CO_2 capture and combustion efficiency". In: *International Journal of Greenhouse Gas Control* 50 (2016), pp. 80–92.

[27] Pérez-Vega, R., Adánez-Rubio, I., Gayán, P., Izquierdo, M. T., Abad, A., García-Labiano, F., de Diego, L. F., and Adánez, J. "Sulphur, nitrogen and mercury emissions from coal combustion with CO_2 capture in chemical looping with oxygen uncoupling (CLOU)". In: *International Journal of Greenhouse Gas Control* 46 (2016), pp. 28–38.

[28] Leion, H., Mattisson, T., and Lyngfelt, A. "The use of petroleum coke as fuel in chemical-looping combustion". In: *Fuel* 86.12-13 (2007), pp. 1947–1958.

[29] Linderholm, C., Schmitz, M., Knutsson, P., and Lyngfelt, A. "Chemical-looping combustion in a 100-kW unit using a mixture of ilmenite and manganese ore as oxygen carrier". In: *Fuel* 166 (2016), pp. 533–542.

[30] Linderholm, C. and Schmitz, M. "Chemical-looping combustion of solid fuels in a 100 kW dual circulating fluidized bed system using iron ore as oxygen carrier". In: *Journal of Environmental Chemical Engineering* 4.1 (2016), pp. 1029–1039.

[31] Mattisson, T., Leion, H., and Lyngfelt, A. "Chemical-looping with oxygen uncoupling using CuO/ZrO_2 with petroleum coke". In: *Fuel* 88.4 (2009), pp. 683–690.

[32] Moldenhauer, P., Rydén, M., Mattisson, T., Hoteit, A., Jamal, A., and Lyngfelt, A. "Chemical-looping combustion with fuel oil in a 10 kW pilot plant". In: *Energy and Fuels* 28.9 (2014), pp. 5978–5987.

[33] Ge, H., Shen, L., Gu, H., Song, T., and Jiang, S. "Combustion performance and sodium transformation of high-sodium ZhunDong coal during chemical looping combustion with hematite as oxygen carrier". In: *Fuel* 159 (2015), pp. 107–117.

[34] Gu, H., Shen, L., Zhong, Z., Zhou, Y., Liu, W., Niu, X., Ge, H., Jiang, S., and Wang, L. "Interaction between biomass ash and iron ore oxygen carrier during chemical looping combustion". In: *Chemical Engineering Journal* 277 (2015), pp. 70–78.

[35] Gu, H., Shen, L., Zhang, S., Niu, M., Sun, R., and Jiang, S. "Enhanced fuel conversion by staging oxidization in a continuous chemical looping reactor based on iron ore oxygen carrier". In: *Chemical Engineering Journal* 334.July 2017 (2018), pp. 829–836.

[36] Shen, L., Wu, J., Gao, Z., and Xiao, J. "Reactivity deterioration of NiO/Al_2O_3 oxygen carrier for chemical looping combustion of coal in a 10 kW_{th} reactor". In: *Combustion and Flame* 156.7 (2009), pp. 1377–1385.

[37] Shen, L., Wu, J., Xiao, J., Song, Q., and Xiao, R. "Chemical-looping combustion of biomass in a 10 kW_{th} reactor with iron oxide as an oxygen carrier". In: *Energy and Fuels* 23.5 (2009), pp. 2498–2505.

[38] Shen, L., Wu, J., Gao, Z., and Xiao, J. "Characterization of chemical looping combustion of coal in a 1 kW_{th} reactor with a nickel-based oxygen carrier". In: *Combustion and Flame* 157.5 (2010), pp. 934–942.

[39] Song, T., Wu, J., Zhang, H., and Shen, L. "Characterization of an Australia hematite oxygen carrier in chemical looping combustion with coal". In: *International Journal of Greenhouse Gas Control* 11 (2012), pp. 326–336.

[40] Song, T., Shen, T., Shen, L., Xiao, J., Gu, H., and Zhang, S. "Evaluation of hematite oxygen carrier in chemical-looping combustion of coal". In: *Fuel* 104 (2013), pp. 244–252.

[41] Xiao, R., Chen, L., Saha, C., Zhang, S., and Bhattacharya, S. "Pressurized chemical-looping combustion of coal using an iron ore as oxygen carrier in a pilot-scale unit". In: *International Journal of Greenhouse Gas Control* 10 (2012), pp. 363–373.

[42] Song, T., Hartge, E. U., Heinrich, S., Shen, L., and Werther, J. "Chemical looping combustion of high sodium lignite in the fluidized bed: Combustion performance and sodium transfer". In: *International Journal of Greenhouse Gas Control* 70 (2018), pp. 22–31.

[43] Thon, A. "Operation of a system of interconnected fluidized bed reactors in the chemical looping combustion process". PhD thesis. Technische Universität Hamburg-Harburg, 2013.

[44] Kramp, M., Thon, A., Hartge, E. U., Heinrich, S., and Werther, J. "Carbon Stripping – A Critical Process Step in Chemical Looping Combustion of Solid Fuels". In: *Chemical Engineering and Technology* 35.3 (2012), pp. 497–507.

[45] Bao, J., Li, Z., and Cai, N. "Experiments of char particle segregation effect on the gas conversion behavior in the fuel reactor for chemical looping combustion". In: *Applied Energy* 113 (2014), pp. 1874–1882.

[46] Cheng, M., Li, Y., Li, Z., and Cai, N. "An integrated fuel reactor coupled with an annular carbon stripper for coal-fired chemical looping combustion". In: *Powder Technology* 320 (2017), pp. 519–529.

[47] Ma, J., Zhao, H., Tian, X., Wei, Y., Rajendran, S., Zhang, Y., Bhattacharya, S., and Zheng, C. "Chemical looping combustion of coal in a 5 kW$_{th}$ interconnected fluidized bed reactor using hematite as oxygen carrier". In: *Applied Energy* 157 (2015), pp. 304–313.

[48] Yang, W., Zhao, H., Wang, K., and Zheng, C. "Synergistic effects of mixtures of iron ores and copper ores as oxygen carriers in chemical-looping combustion". In: *Proceedings of the Combustion Institute* 35.3 (2015), pp. 2811–2818.

[49] Ströhle, J., Orth, M., and Epple, B. "Design and operation of a 1 MW$_{th}$ chemical looping plant". In: *Applied Energy* 113 (2014), pp. 1490–1495.

[50] Kolbitsch, P., Bolhàr-Nordenkampf, J., Pröll, T., and Hofbauer, H. "Operating experience with chemical looping combustion in a 120 kW dual

circulating fluidized bed (DCFB) unit". In: *International Journal of Greenhouse Gas Control* 4.2 (2010), pp. 180–185.

[51] Marx, K., Bolhàr-Nordenkampf, J., Pröll, T., and Hofbauer, H. "Chemical looping combustion for power generation – Concept study for a 10 MW_{th} demonstration plant". In: *International Journal of Greenhouse Gas Control* 5.5 (2011), pp. 1199–1205.

[52] Pröll, T., Mayer, K., Bolhàr-Nordenkampf, J., Kolbitsch, P., Mattisson, T., Lyngfelt, A., and Hofbauer, H. "Natural minerals as oxygen carriers for chemical looping combustion in a dual circulating fluidized bed system". In: *Energy Procedia* 1.1 (2009), pp. 27–34.

[53] Yin, S., Shen, L., Dosta, M., Hartge, E. U., Heinrich, S., Lu, P., Werther, J., and Song, T. "Chemical Looping Gasification of a Biomass Pellet with a Manganese Ore as an Oxygen Carrier in the Fluidized Bed". In: *Energy and Fuels* 32.11 (2018), pp. 11674–11682.

[54] Acosta-Iborra, A., Hernández-Jiménez, F., Sobrino, C., and Vega, M. de. "Experimental and computational study on the bubble behavior in a 3-D fluidized bed with a vertical-axis, rotating distributor". In: *The 13th International Conference on Fluidization - New Paradigm in Fluidization Engineering*. Gyeong-ju, Korea, 2010.

[55] Werther, J. and Molerus, O. "The local structure of gas fluidized beds – II. the spatial distribution of bubbles". In: *Int. J. Multiphase Flow* 1.1 (1973), pp. 123–138.

[56] Werther, J. and Molerus, O. "Autokorrelation und Kreuzkorrelation zur Messung lokaler Blasengrößen und -aufstiegsgeschindigkeiten in realen Gas-Feststoff-Fließbetten". In: *Chemie-Ing.-Techn.* 43.5 (1971), pp. 271–273.

[57] Lim, C. N., Gilbertson, M. A., and Harrison, A. J. "Bubble distribution and behaviour in bubbling fluidised beds". In: *Chemical Engineering Science* 62.1-2 (2007), pp. 56–69.

[58] Penn, A., Boyce, C. M., Pruessmann, K. P., and Müller, C. R. "Regimes of jetting and bubbling in a fluidized bed studied using real-time magnetic resonance imaging". In: *Chemical Engineering Journal* 383.123185 (2020).

[59] Davidson, J. F. and Harrison, D. *Fluidised particles*. 1st. Cambridge: Cambridge University Press, 1963.

[60] Boyce, C. M., Penn, A., Lehnert, M., Pruessmann, K. P., and Müller, C. R. "Magnetic resonance imaging of single bubbles injected into incipiently fluidized beds". In: *Chemical Engineering Science* 200 (2019), pp. 147–166.

[61] Yasui, G. and Johanson, L. N. "Characteristics of gas pockets in fluidized beds". In: *AIChE Journal* 4.4 (1958), pp. 445–452.

[62] Kai, T., Misawa, M., Takahashi, T., Tiseanu, I., and Ichikawa, N. "Observation of 3-D structure of bubbles in a fluidized catalyst bed". In: *The Canadian Journal of Chemical Engineering* 83.1 (2005), pp. 113–118.

[63] Darton, R. C., LaNauze, R. D., Davidson, J. F., and Harrison, D. "Bubble growth due To coalescence in fluidised beds". In: *Trans Inst Chem Eng* 55 (1977), pp. 274–280.

[64] van Lare, C. E., Piepers, H. W., Schoonderbeek, J. N., and Thoenes, D. "Investigation on bubble characteristics in a gas fluidized bed". In: *Chemical Engineering Science* 52.5 (1997), pp. 829–841.

[65] Shen, L., Johnsson, F., and Leckner, B. "Digital image analysis of hydrodynamics two-dimensional bubbling fluidized beds". In: *Chemical Engineering Science* 59.13 (2004), pp. 2607–2617.

[66] Geldart, D. "The size and frequency of bubbles in two- and three-dimensional gas-fluidised beds". In: *Powder Technology* 4.1 (1970), pp. 41–55.

[67] Hilligardt, K. and Werther, J. "Influence of temperature and properties of solids on the size and growth of bubbles in gas fluidized beds". In: *Chemical Engineering & Technology* 10.1 (1987), pp. 272–280.

[68] Choi, J. H., Son, J. E., and Kim, S. D. "Bubble size and frequency in gas fluidized beds". In: *Journal of Chemical Engineering of Japan* 21.2 (1988), pp. 171–178.

[69] Mostoufi, N. and Chaouki, J. "Flow structure of the solids in gas-solid fluidized beds". In: *Chemical Engineering Science* 59.20 (2004), pp. 4217–4227.

[70] Abrahamsen, A. R. and Geldart, D. "Behaviour of gas-fluidized beds of fine powders part I. Homogeneous expansion". In: *Powder Technology* 26.1 (1980), pp. 35–46.

[71] Chen, H., Yang, D., and Cheng, J. "Hydrodynamics of gas solids in a bubbling fluidized bed with binary particles". In: *Procedia Engineering* 102 (2015), pp. 799–803.

[72] Shabanian, J. and Chaouki, J. "Effects of temperature, pressure, and interparticle forces on the hydrodynamics of a gas-solid fluidized bed". In: *Chemical Engineering Journal* 313 (2017), pp. 580–590.

[73] Čárský, M., Hartman, M., Ilyenko, B. K., and Makhorin, K. E. "The bubble frequency in a fluidized bed at elevated pressure". In: *Powder Technology* 61.3 (1990), pp. 251–254.

[74] Huang, J. and Lu, Y. "Characteristics of bubble, cloud and wake in jetting fluidised bed determined using a capacitance probe". In: *Chemical Engineering Research and Design* 136 (2018), pp. 687–697.

[75] Grace, J. R. and Harrison, D. "The behaviour of freely bubbling fluidised beds". In: *Chemical Engineering Science* 24.3 (1969), pp. 497–508.

[76] Groen, J. S., Oldemam, R. G., Mudde, R. F., and van den Akker, H. E. "Coherent structures and axial dispersion in bubble column reactors". In: *Chemical Engineering Science* 51.10 (1996), pp. 2511–2520.

[77] Davidson, J. F. and Harrison, D. "The behaviour of a continuously bubbling fluidised bed". In: *Chemical Engineering Science* 21.9 (1966), pp. 731–738.

[78] Puncochar, M., Ruzicka, M. C., and Simcik, M. "Bubble swarm rise velocity in fluidized beds". In: *Chemical Engineering Science* 152 (2016), pp. 84–94.

[79] Werther, J. "Bubble chains in large diameter gas fluidized beds". In: *International Journal of Multiphase Flow* 3.4 (1977), pp. 367–381.

[80] Boyce, C. M., Penn, A., Lehnert, M., Pruessmann, K. P., and Müller, C. R. "Wake volume of injected bubbles in fluidized beds: A magnetic resonance imaging velocimetry study". In: *Powder Technology* 357 (2019), pp. 428–435.

[81] Kunii, D. and Levenspiel, O. *Fluidization Engineering*. 2nd ed. Boston: Butterworth-Heinemann, 1991.

[82] Gera, D. and Gautam, M. "Effect of bubble coalescence on throughflow velocity in a 2-D fluidized bed". In: *Powder Technology* 83.1 (1995), pp. 49–53.

[83] Penn, A., Boyce, C. M., Kovar, T., Tsuji, T., Pruessmann, K. P., and Müller, C. R. "Real-Time Magnetic Resonance Imaging of Bubble Behavior and Particle Velocity in Fluidized Beds". In: *Industrial and Engineering Chemistry Research* 57.29 (2018), pp. 9674–9682.

[84] Boyce, C. M., Rice, N. P., Ozel, A., Davidson, J. F., Sederman, A. J., Gladden, L. F., Sundaresan, S., Dennis, J. S., and Holland, D. J. "Magnetic resonance characterization of coupled gas and particle dynamics in a bubbling fluidized bed". In: *Physical Review Fluids* 1.7 074201 (2016).

[85] Kobayashi, N., Yamazaki, R., and Mori, S. "A study on the behavior of bubbles and solids in bubbling fluidized beds". In: *Powder Technology* 113.3 (2000), pp. 327–344.

[86] Sette, E., García, A. G., Pallarès, D., and Johnsson, F. "Quantitative Evaluation of Inert Solids Mixing in a Bubbling Fluidized Bed". In: *21st International conference on fluidized bed combustion*. Naples, Italy, 2012.

[87] Chew, J. W., Wolz, J. R., and Hrenya, C. M. "Axial Segregation in Bubbling Gas-Fluidized Beds with Gaussian and Lognormal Distributions of Geldart Group B Particles". In: *AICHE Journal* 56.12 (2010), pp. 3046–3061.

[88] Joseph, G. G., Leboreiro, J., Hrenya, C. M., and Stevens, A. R. "Experimental segregation profiles in bubbling gas-fluidized beds". In: *AIChE Journal* 53.11 (2007), pp. 2804–2813.

[89] Rowe, P. N. and Matsuno, R. "Single bubbles injected into a gas fluidised bed and observed by X-rays". In: *Chemical Engineering Science* 26.6 (1971), pp. 923–935.

[90] Dang, N., Kolkman, T., Gallucci, F., and van Sint Annaland, M. "Development of a Novel Infrared Camera for Gas Exchange from Bubble-to-Emulsion Phase in Gas-Solid Fluidized Beds". In: *The 14th International Conference on Fluidization*. Noordwijkerhout, The Netherlands, 2013.

[91] van Deemter, J. J. "Mixing". In: *Fluidization*. Ed. by Davidson, J. F., Clift, R., and Harrison, D. London, New York: Academic Press, 1985. Chap. 9.

[92] Johnsson, F., Andersson, S., and Leckner, B. "Expansion of a freely bubbling fluidized bed". In: *Powder Technology* 68.2 (1991), pp. 117–123.

[93] Geldart, D. "Expansion of gas fluidized beds". In: *Industrial and Engineering Chemistry Research* 43.18 (2004), pp. 5802–5809.

[94] Wiman, J. and Almstedt, A. E. "Influence of pressure, fluidization velocity and particle size on the hydrodynamics of a freely bubbling fluidized bed". In: *Chemical Engineering Science* 53.12 (1998), pp. 2167–2176.

[95] Formisani, B., Girimonte, R., and Pataro, G. "The influence of operating temperature on the dense phase properties of bubbling fluidized beds of solids". In: *Powder Technology* 125.1 (2002), pp. 28–38.

[96] Chan, C., Seville, J. P., and Baeyens, J. "The transport disengagement height (TDH) in a bubbling fluidized bed". In: *The 13th International Conference on Fluidization - New Paradigm in Fluidization Engineering* (2010), pp. 1–8.

[97] Holland, D. J., Müller, C. R., Dennis, J. S., Gladden, L. F., and Davidson, J. F. "Magnetic Resonance Studies of Fluidization Regimes". In: *Industrial & Engineering Chemistry Research* 49.12 (2010), pp. 5891–5899.

[98] Bi, H. T. and Su, P. C. "Local phase holdups in gas-solids fluidization and transport". In: *AIChE Journal* 47.9 (2001), pp. 2025–2031.

[99] Wytrwat, T., Hartge, E.-U., Yazdanpanah, M., and Heinrich, S. "Capacitance probes for the investigation of the fluid dynamics of turbulent fluidized beds". In: *12th International Conference on Fluidized Bed Technology, CFB 2017*. Krakow, Poland, 2017, pp. 245–252.

[100] Qi, M., Barghi, S., and Zhu, J. "Detailed hydrodynamics of high flux gas-solid flow in a circulating turbulent fluidized bed". In: *Chemical Engineering Journal* 209 (2012), pp. 633–644.

[101] Ellis, N. "Hydrodynamics of Gas-Solid Turbulent Fluidized Beds". PhD thesis. The University of British Columbia, 2003.

[102] Chaouki, J., Gonzalez, A., Guy, C., and Klvana, D. "Two-phase model for a catalytic turbulent fluidized-bed reactor: Application to ethylene synthesis". In: *Chemical Engineering Science* 54.13-14 (1999), pp. 2039–2045.

[103] Du, B., Warsito, W., and Fan, L. S. "Bed nonhomogeneity in turbulent gas-solid fluidization". In: *AIChE Journal* 49.5 (2003), pp. 1109–1126.

[104] Ellis, N., Bi, H. T., Lim, C. J., and Grace, J. R. "Hydrodynamics of turbulent fluidized beds of different diameters". In: *Powder Technology* 141.1-2 (2004), pp. 124–136.

[105] Zhu, H., Zhu, J., Li, G., and Li, F. "Detailed measurements of flow structure inside a dense gas-solids fluidized bed". In: *Powder Technology* 180.3 (2008), pp. 339–349.

[106] Bi, H. and Fan, L.-S. .-S. "Existence of turbulent regime in gas-solid fluidization". In: *AIChE Journal* 38.2 (1992), pp. 297–301.

[107] Clift, R., Grace, J. R., and Weber, M. E. "Stability of Bubbles in Fluidized Beds". In: *Industrial and Engineering Chemistry Fundamentals* 13.1 (1974), pp. 45–51.

[108] Andreux, R., Gauthier, T., Chaouki, J., and Simonin, O. "New description of fluidization regimes". In: *AIChE Journal* 51.4 (2005), pp. 1125–1130.

[109] Lancia, A., Nigro, R., Volpicelli, G., and Santoro, L. "Transition from slugging to turbulent flow regimes in fluidized beds detected by means of capacitance probes". In: *Powder Technology* 56.1 (1988), pp. 49–56.

[110] Werther, J. and Wein, J. "Expansion Behavior of Gas Fluidized Beds in the Turbulent Regime". In: *AIChE Symposium Series* 90.301 (1994), pp. 31–44.

[111] Seo, M. W., Goo, J. H., Kim, S. D., Lee, J. G., Guahk, Y. T., Rho, N. S., Koo, G. H., Lee, D. Y., Cho, W. C., and Song, B. H. "The transition velocities in a dual circulating fluidized bed reactor with variation of temperatures". In: *Powder Technology* 264 (2014), pp. 583–591.

[112] Wytrwat, T., Hartge, E.-U., Yazdanpanah, M., and Heinrich, S. "Transition from Bubbling to Turbulent Fluidization for Geldart's Group B Particles". In: *23rd International Conference on FBC*. Seoul, Korea, 2018, pp. 642–651.

[113] Cai, P., Chen, S. P., Jin, Y., Yu, Z. Q., and Wang, Z. W. "Effect of Operating Temperature and Pressure on the Transition from Bubbling to Turbulent Fluidization". In: *AIChE Symposium Series* 85.270 (1989), pp. 37–43.

[114] Chehbouni, A., Chaouki, J., Guy, C., and Klvana, D. "Effect of Temperature on the Hydrodynamics of Turbulent Fluidized Beds". In: *Fluidization VIII*. Ed. by Large, J.-F. and Laguérie, C. Tours, France, 1995, pp. 149–156.

[115] Rim, G. H. and Lee, D. H. "Bubbling to turbulent bed regime transition of ternary particles in a gas-solid fluidized bed". In: *Powder Technology* 290 (2016), pp. 45–52.

[116] Peeler, P. K., Lim, K. S., and Close, R. C. "Effect of temperature on the turbulent fluidization regime transition". In: *6th International conference on circulating fluidized beds*. Würzburg, Germany, 1999, pp. 125–130.

[117] Hagh-Shenas-Lari, M. J. and Mostoufi, N. "Effect of temperature on fluidization regimes". In: *Chemical Engineering and Technology* 37.9 (2014), pp. 1593–1599.

[118] Yazdanpanah, M., Forret, A., and Gauthier, T. "Impact of size and temperature on the hydrodynamics of chemical looping combustion". In: *Applied Energy* 157 (2015), pp. 416–421.

[119] Bi, H. T. and Grace, J. R. "Effect of measurement method on the velocities used to demarcate the onset of turbulent fluidization". In: *The Chemical Engineering Journal and The Biochemical Engineering Journal* 57.3 (1995), pp. 261–271.

[120] Grace, J. R. and Sun, G. "Influence of particle size distribution on the performance of fluidized bed reactors". In: *The Canadian Journal of Chemical Engineering* 69.5 (1991), pp. 1126–1134.

[121] Sun, G. and Grace, J. R. "Effect of particle size distribution in different fluidization regimes". In: *AIChE Journal* 38.5 (1992), pp. 716–722.

[122] Zhang, X. and Bi, H. T. "Study on Void Behavior in a Turbulent Fluidized Bed with Catalyst Powders". In: *Industrial & Engineering Chemistry Research* 49.15 (2010), pp. 6862–6869.

[123] Wytrwat, T., Yazdanpanah, M., and Heinrich, S. "Bubble Properties in Bubbling and Turbulent Fluidized Beds for Particles of Geldart's Group B". In: *Processes* 8.9 (2020), p. 1098.

[124] Horio, M. and Nonaka, A. "A generalized bubble diameter correlation for gas-solid fluidized beds". In: *AIChE Journal* 33.11 (1987), pp. 1865–1872.

[125] Lee, G. S. and Kim, S. D. "Void characteristics in turbulent fluidized beds". In: *Korean Journal of Chemical Engineering* 8.1 (1991), pp. 6–11.

[126] Ellis, N., Briens, L. A., Grace, J. R., Bi, H. T., and Lim, C. J. "Characterization of dynamic behaviour in gas-solid turbulent fluidized bed using chaos and wavelet analyses". In: *Chemical Engineering Journal* 96.1-3 (2003), pp. 105–116.

[127] Sobrino, C., Ellis, N., and De Vega, M. "Distributor effect in the bottom region of turbulent fluidized bed: Implications to scale-up". In: *CFB 2008 - Proceedings of the 9th Int. Conference on Circulating Fluidized Beds, in Conjunction with the 4th International VGB Workshop "Operating Experience with Fluidized Bed Firing Systems"*. Vol. c. Hamburg, Germany, 2008.

[128] Lee, G. S. and Kim, S. D. "Bed expansion characteristics and transition velocity in turbulent fluidized beds". In: *Powder Technology* 62.3 (1990), pp. 207–215.

[129] Zhu, H. and Zhu, J. "Comparative study of flow structures in a circulating-turbulent fluidized bed". In: *Chemical Engineering Science* 63.11 (2008), pp. 2920–2927.

[130] Qi, X., Zhu, H., and Zhu, J. "Demarcation of a new circulating turbulent fluidization regime". In: *AIChE Journal* 55.3 (2009), pp. 594–611.

[131] Zhu, J. "Circulating turbulent fluidization – A new fluidization regime or just a transitional phenomenon". In: *Particuology* 8.6 (2010), pp. 640–644.

[132] Qi, M., Zhu, J., and Barghi, S. "Particle velocity and flux distribution in a high solids concentration circulating turbulent fluidized bed". In: *Chemical Engineering Science* 84 (2012), pp. 437–448.

[133] Richardson, J. F. and Zaki, W. N. "Sedimentation and fluidization: Part I". In: *Process Safety and Environmental Protection: Transactions of the Institution of Chemical Engineers, Part B* 75.Supplement (1997), S82–S100.

[134] Adánez, J., Gayán, P., García-Labiano, F., and de Diego, L. "Axial voidage profiles in fast fluidized beds". In: *Powder Technology* 81.3 (1994), pp. 259–268.

[135] Lei, H. and Horio, M. "A Comprehensive Pressure Balance Model of Circulating Fluidized Beds". In: *Journal of Chemical Engineering of Japan* 31.1 (1998), pp. 83–94.

[136] Wen, C. Y. and Chen, L. H. "Fluidized bed freeboard phenomena: Entrainment and elutriation". In: *AIChE Journal* 28.1 (1982), pp. 117–128.

[137] Rhodes, M. and Geldart, D. "A model for the circulating fluidized bed". In: *Powder Technology* 53.3 (1987), pp. 155–162.

[138] Kunii, D. and Levenspiel, O. "Entrainment of solids from fluidized beds I. Hold-up of solids in the freeboard II. Operation of fast fluidized beds". In: *Powder Technology* 61.2 (1990), pp. 193–206.

[139] Cui, H., Mostoufi, N., and Chaouki, J. "Gas and solids between dynamic bubble and emulsion in gas-fluidized beds". In: *Powder Technology* 120.1-2 (2001), pp. 12–20.

[140] Lee, G. S. and Kim, S. D. "Axial mixing of solids in turbulent fluidized beds". In: *The Chemical Engineering Journal* 44.1 (1990), pp. 1–9.

[141] Lee, G. S. and Kim, S. D. "Gas Mixing in Slugging and Turbulent Fluidized Beds". In: *Chemical Engineering Communications* 86.1 (1989), pp. 91–111.

[142] Zhang, Y., Lu, C., Grace, J. R., Bi, X., and Shi, M. "Gas Back-Mixing in a Two-Dimensional Baffled Turbulent Fluidized Bed". In: *Industrial & Engineering Chemistry Research* 47.21 (2008), pp. 8484–8491.

[143] Foka, M., Chaouki, J., Guy, C., and Klvana, D. "Gas phase hydrodynamics of a gas-solid turbulent fluidized bed reactor". In: *Chemical Engineering Science* 51.5 (1996), pp. 713–723.

[144] Zhang, Y., Fan, Y., Lu, C., and Shi, M. "Gas backmixing in a two-dimensional baffled turbulent fluidized bed". In: *CFB 2008 - Proceedings of the 9th Int. Conference on Circulating Fluidized Beds, in Conjunction with the 4th International VGB Workshop "Operating Experience with Fluidized Bed Firing Systems"*. Hamburg, Germany, 2008.

[145] Thompson, M. L., Bi, H., and Grace, J. R. "A generalized bubbling/turbulent fluidized-bed reactor model". In: *Chemical Engineering Science* 54.13-14 (1999), pp. 2175–2185.

[146] Miyauchi, T., Furusaki, S., Yamada, K., and Matsumura, M. "Experimental Determinations of the Vertical Distribution of Contact Efficiency Inside a Fluidized Catalyst Bed". In: *Fluidization*. Boston, MA: Springer US, 1980, pp. 571–580.

[147] Fotovat, F. "Entrainment from Bubbling and Turbulent Beds". In: *Essentials of Fluidization Technology*. Weinheim: Wiley-VCH Verlag GmbH & Co. KGaA, 2020, pp. 181–202.

[148] Werther, J. and Hartge, E.-U. "Elutriation and Entrainment". In: *Handbook of Fluidization and Fluid-Particle Systems*. Ed. by Yang, W.-C. 1st ed. New York, Basel: Marcel Dekker, Inc., 2003, pp. 113–128.

[149] Tasirin, S. and Geldart, D. "Entrainment of FCC from fluidized beds — A new correlation for the elutriation rate constants $K_{i\infty}^*$". In: *Powder Technology* 95.3 (1998), pp. 240–247.

[150] van Swaaij, W. P. "The Design of Gas-Solids Fluid Bed and Related Reactors". In: *ACS Symposium Series*. Vol. 72. Chapter 6. 1978, pp. 193–222.

[151] Bos, R. A. N., Tromp, P. J., and Akse, H. N. "Conversion of Methanol to Lower Olefins. Kinetic Modeling, Reactor Simulation, and Selection". In:

Industrial and Engineering Chemistry Research 34.11 (1995), pp. 3808–3816.

[152] Edwards, M. and Avidan, A. "Conversion model aids scale-up of Mobil's fluid-bed MTG process". In: *Chemical Engineering Science* 41.4 (1986), pp. 829–835.

[153] Foka, M., Chaouki, J., Guy, C., and Klvana, D. "Natural gas combustion in a catalytic turbulent fluidized bed". In: *Chemical Engineering Science* 49.24 Part A (1994), pp. 4269–4276.

[154] Muradov, N., Chen, Z., and Smith, F. "Fossil hydrogen with reduced CO2 emission: Modeling thermocatalytic decomposition of methane in a fluidized bed of carbon particles". In: *International Journal of Hydrogen Energy* 30.10 (2005), pp. 1149–1158.

[155] Krambeck, F. J., Avidan, A. A., Lee, C. K., and Lo, M. N. "Predicting fluid-bed reactor efficiency using adsorbing gas tracers". In: *AIChE Journal* 33.10 (1987), pp. 1727–1734.

[156] Sun, G. and Grace, J. R. "The effect of particle size distribution on the performance of a catalytic fluidized bed reactor". In: *Chemical Engineering Science* 45.8 (1990), pp. 2187–2194.

[157] Venderbosch, R. H. "The role of clusters in gas-solids reactors. An experimental study". PhD thesis. Twente University, 1998.

[158] Mostoufi, N., Cui, H., and Chaouki, J. "A comparison of two- and single-phase models for fluidized-bed reactors". In: *Industrial and Engineering Chemistry Research* 40.23 (2001), pp. 5526–5532.

[159] Eslami, A., Hashemi Sohi, A., Sheikhi, A., and Sotudeh-Gharebagh, R. "Sequential modeling of coal volatile combustion in fluidized bed reactors". In: *Energy and Fuels* 26.8 (2012), pp. 5199–5209.

[160] Ege, P., Grislingås, A., and De Lasa, H. I. "Modelling turbulent fluidized bed reactors: Tracer and fibre optic probe studies". In: *Chemical Engineering Journal and the Biochemical Engineering Journal* 61.3 (1996), pp. 179–190.

[161] Farag, H. I., Ege, P. E., Grislingås, A., and De Lasa, H. I. "Flow Patterns in a Pilot Plant-Scale Turbulent Fluidized Bed Reactor: Concurrent Application of Tracers and Fiber Optic Sensors". In: *Canadian Journal of Chemical Engineering* 75.5 (1997), pp. 851–860.

[162] Sotudeh-Gharebaagh, R. and Chaouki, J. "Gas mixing in a turbulent fluidized bed reactor". In: *The Canadian Journal of Chemical Engineering* 78.1 (2000), pp. 65–74.

[163] Sotudeh-Gharebaagh, R. and Chaouki, J. "Investigation of highly exothermic reactions in a turbulent fluidized bed reactor". In: *Energy and Fuels* 21.4 (2007), pp. 2230–2237.

[164] Abba, I. A., Grace, J. R., Bi, H. T., and Thompson, M. L. "Spanning the flow regimes: Generic fluidized-bed reactor model". In: *AIChE Journal* 49.7 (2003), pp. 1838–1848.

[165] Guo, F. "Gas flow and mixing behavior in fine-powder fluidized bed". In: *AIChE Journal* 33.11 (1987), pp. 1895–1898.

[166] Jafari, R., Sotudeh-Gharebagh, R., and Mostoufi, N. "Performance of the wide-ranging models for fluidized bed reactors". In: *Advanced Powder Technology* 15.5 (2004), pp. 533–548.

[167] Pugsley, T. S. and Berruti, F. "A predictive hydrodynamic model for circulating fluidized bed risers". In: *Powder Technology* 89.1 (1996), pp. 57–69.

[168] Wytrwat, T., Hartge, E.-U., Yazdanpanah, M., and Heinrich, S. "Influences on the transition from bubbling to turbulent fluidization for Geldart's group B particles". In: *Powder Technology* 375 (2020), pp. 81–88.

[169] DIN EN ISO 5167-2:2003, German Institute for Standardization - Fundamental Technical Standards Committee. *Measurement of fluid flow by means of pressure differential devices inserted in circular cross-section conduits running full - Part 2: Orifice plates.*

[170] ISO 13322-2:2006-11, ISO/TC 24 Sieves; sieving and other sizing methods. *Particle size analysis - Image analysis methods - Part 2: Dynamic image analysis methods.*

[171] DIN EN ISO 60:2000-01, German Institute for Standardization - Plastics Standards Committee. *Bestimmung der scheinbaren Dichte von Formmassen, die durch einen genormten Trichter abfließen können (Schüttdichte).*

[172] Cheng, N.-S. "Comparison of formulas for drag coefficient and settling velocity of spherical particles". In: *Powder Technology* 189.3 (2009), pp. 395–398.

[173] Savitzky, A. and Golay, M. J. E. "Smoothing and Differentiation of Data by Simplified Least Squares Procedures." In: *Analytical Chemistry* 36.8 (1964), pp. 1627–1639.

[174] Kage, H., Agari, M., Ogura, H., and Matsuno, Y. "Frequency analysis of pressure fluctuation in fluidized bed plenum and its confidence limit for detection of various modes of fluidization". In: *Advanced Powder Technology* 11.4 (2000), pp. 459–475.

[175] van der Schaaf, J., van Ommen, J. R., Takens, F., Schouten, J. C., and van den Bleek, C. M. "Similarity between chaos analysis and frequency analysis of pressure fluctuations in fluidized beds". In: *Chemical Engineering Science* 59.8-9 (2004), pp. 1829–1840.

[176] van Ommen, J. R., Sasic, S., van der Schaaf, J., Gheorghiu, S., Johnsson, F., and Coppens, M. O. "Time-series analysis of pressure fluctuations in gas-solid fluidized beds – A review". In: *International Journal of Multiphase Flow* 37.5 (2011), pp. 403–428.

[177] He, H., Lu, X., Shuang, W., Wang, Q., Kang, Y., Yan, L., Ji, X., Luo, G., and Liu, H. "Statistical and frequency analysis of the pressure fluctuation in a fluidized bed of non-spherical particles". In: *Particuology* 16 (2014), pp. 178–186.

[178] Krishnan, V. "Random Processes". In: *Probability and Random Processes*. Vol. 22. 8. Hoboken, NJ, USA: John Wiley & Sons, Inc., 2005, pp. 406–489.

[179] Thomopoulos, N. T. "Normal". In: *Statistical Distributions*. Cham: Springer International Publishing, 2017, pp. 69–76.

[180] Werther, J. "Measurement techniques in fluidized beds". In: *Powder Technology* 102.1 (1999), pp. 15–36.

[181] Geldart, D. and Kelsey, J. R. "The use of capacitance probes in gas fluidised beds". In: *Powder Technology* 6.1 (1972), pp. 45–50.

[182] Werther, J. and Molerus, O. "The local structure of gas fluidized beds – I. A statistically based measuring system". In: *International Journal of Multiphase Flow* 1.1 (1973), pp. 103–122.

[183] Hage, B. and Werther, J. "The guarded capacitance probe – A tool for the measurement of solids flow patterns in laboratory and industrial fluidized bed combustors". In: *Powder Technology* 93.3 (1997), pp. 235–245.

[184] Wiesendorf, V. and Werther, J. "Capacitance probes for solids volume concentration and velocity measurements in industrial fluidized bed reactors". In: *Powder Technology* 110.1-2 (2000), pp. 143–157.

[185] Lancia, A., Nigro, R., Santoro, L., and Volpicelli, G. "Characterization of the quality of slugging regime in fluidized beds by means of two capacitance probes". In: *Advanced Powder Technology* 2.1 (1991), pp. 37–47.

[186] Farag, H. I., Mejdell, T., Hjarbo, K., Ege, P., Lysberg, M., Grislingås, A., and De Lasa, H. "Fibre optic and capacitance probes in turbulent fluidized beds". In: *Chemical Engineering Communications* 157.1 (1997), pp. 73–107.

[187] Wiesendorf, V. "The capacitance probe – a tool for flow investigations in gas-solids fluidization systems". PhD thesis. Technische Universität Hamburg-Harburg, 2000.

[188] Stocker, T. C. and Steinke, I. *Statistik: Grundlagen und Methodik.* Berlin, Boston: De Gruyter Oldenbourg, 2016.

[189] Pesch, S. J. "Experimental Analysis and Modeling of Oil Droplet Formation and Rise Behavior under Deep-Sea Conditions". PhD thesis. Hamburg University of Technology, 2020.

[190] Forbes, C., Evans, M., Hastings, N., and Peacock, B. "Lognormal Distribution". In: *Statistical Distributions.* Hoboken, NJ, USA: John Wiley & Sons, Inc., 2010, pp. 131–134.

[191] Flemmer, R. L., Pickett, J., and Clark, N. N. "An experimental study on the effect of particle shape on fluidization behavior". In: *Powder Technology* 77.2 (1993), pp. 123–133.

[192] Wu, S. and Baeyens, J. "Effect of operating temperature on minimum fluidization velocity". In: *Powder Technology* 67.2 (1991), pp. 217–220.

[193] Yang, T. Y. and Leu, L. p. "Study of transition velocities from bubbling to turbulent fluidization by statistic and wavelet multi-resolution analysis on absolute pressure fluctuations". In: *Chemical Engineering Science* 63.7 (2008), pp. 1950–1970.

[194] Chehbouni, A., Chaouki, J., Guy, C., and Klvana, D. "Characterization of the Flow Transition between Bubbling and Turbulent Fluidization". In: *Industrial & Engineering Chemistry Research* 33.8 (1994), pp. 1889–1896.

[195] Shabanian, J., Sauriol, P., Rakib, A., and Chaouki, J. "Characterization of gas-solid fluidization at high temperature by analysis of pressure signals". In: *CFB-11: Proceedings of the 11th International Conference on Fluidized Bed Technology.* Beijing, China, 2014, pp. 85–90.

[196] Shabanian, J. and Chaouki, J. "Pressure Signals in a Gas-Solid Fluidized Bed With Thermally Induced Inter-Particle Forces". In: *The 14th Inter-*

national Conference on Fluidization – From Fundamentals to Products. Noordwijkerhout, The Netherlands, 2013.

[197] Sasic, S., Coppens, M.-O., van der Schaaf, J., Gheorghiu, S., Johnsson, F., and van Ommen, J. R. "Dynamics of Gas-Solid Fluidized Beds Through Pressure Fluctuations : A Brief Examination of Analysis". In: *International conference on circulating fluidized beds and fluidized bed technology - CFB10.* Sun River, Oregon, USA, 2011.

[198] Bi, H. T. "A critical review of the complex pressure fluctuation phenomenon in gas-solids fluidized beds". In: *Chemical Engineering Science* 62.13 (2007), pp. 3473–3493.

[199] Lee, G. S. and Kim, S. D. "Pressure fluctuations in turbulent fluidized beds". In: *Journal of Chemical Engineering of Japan* 21.5 (1988), pp. 515–521.

[200] Sasic, S., Leckner, B., and Johnsson, F. "Characterization of fluid dynamics of fluidized beds by analysis of pressure fluctuations". In: *Progress in Energy and Combustion Science* 33.5 (2007), pp. 453–496.

[201] Chyang, C. S. and Lin, Y. C. "Influence of the nature of the roots blower on pressure fluctuations in a fluidized bed". In: *Powder Technology* 127.1 (2002), pp. 19–31.

[202] Brue, E. J. "Pressure Fluctuations as a Diagnostic Tool for Fluidized Beds". PhD thesis. Iowa State University, 1998.

[203] Baskakov, A. P., Tuponogov, V. G., and Filippovsky, N. F. "A study of pressure fluctuations in a bubbling fluidized bed". In: *Powder Technology* 45.2 (1986), pp. 113–117.

[204] Verloop, J. and Heertjes, P. M. "Periodic pressure fluctuations in fluidized beds". In: *Chemical Engineering Science* 29.4 (1974), pp. 1035–1042.

[205] Schlichthärle, P. "Fluid Dynamics and Mixing of Solids and Gas in the Bottom Zone of Circulating Fluidized Beds". PhD thesis. Hamburg University of Technology, 2000.

[206] Kang, I. S. and Leal, L. G. "The drag coefficient for a spherical bubble in a uniform streaming flow". In: *Physics of Fluids* 31.2 (1988), pp. 233–237.

[207] Choi, J. H., Chang, I. Y., Shun, D. W., Yi, C. K., Son, J. E., and Kim, S. D. "Correlation on the particle entrainment rate in gas fluidized beds". In: *Industrial and Engineering Chemistry Research* 38.6 (1999), pp. 2491–2496.

[208] Sciążko, M., Bandrowski, J., and Raczek, J. "On the entrainment of solid particles from a fluidized bed". In: *Powder Technology* 66.1 (1991), pp. 33–39.

[209] Klinzing, G. E. "Dilute-Phase Pneumatic Conveying". In: *Handbook of Fluidization and Fluid-Particle Systems*. Ed. by Yang, W.-C. 1st ed. New York, Basel: Marcel Dekker, Inc., 2003, pp. 619–630.

[210] Sommerfeld, M., Wirth, K.-E., and Muschelknautz, U. "L3 Two-Phase Gas-Solid Flow". In: *VDI Heat Atlas*. Springer Berlin Heidelberg, 2009, pp. 1181–1238.

[211] Hinkle, B. L. "Acceleration of particles and pressure drops encountered in horizontal pneumatic conveying". PhD thesis. Georgia Institute of Technology, 1953.

[212] Monazam, E. R., Panday, R., and Shadle, L. J. "Estimate of solid flow rate from pressure measurement in circulating fluidized bed". In: *Powder Technology* 203.1 (2010), pp. 91–97.

Appendix

Contributing Student Works

In the following student works supervised by the author that contributed to this work are listed:

Thesis type	Author	Year	Title
Bachelor	Steffen Schaper	2019	Untersuchung der lokalen Strömungsstruktur beim Übergang von blasenbildender zu turbulenter Fluidisation für Geldart B Partikel
Master	Lasse Tuchtfeldt	2019	Einfluss der Temperatur auf den Übergang von blasenbildender zu turbulenter Fluidisierung für Partikel der Geldart-Gruppe B
Bachelor	Sinah Kammler	2017	Experimentelle Untersuchung des Übergangs von blasenbildender zu turbulenter Fluidisation für Geldart B Partikel
Bachelor	Labiba Ahmed	2016	Vergleich zweier invasiver Messmethoden zur Charakterisierung der Strömungsmechanik in einer zirkulierenden Wirbelschicht
Bachelor	Helene Heidt	2016	Implementierung eines kapazitiven Messsystems zur Bestimmung der Feststoffverteilung in der Wirbel- und Strahlschicht

Previously Published Publications

This work is partially based on previously published works of the author. These publications are listed in the following.

Peer Reviewed Papers

Wytrwat, T., Yazdanpanah, M., and Heinrich, S. "Bubble Properties in Bubbling and Turbulent Fluidized Beds for Particles of Geldart's Group B". In: *Processes* 8.9 (2020), p. 1098.

Wytrwat, T., Hartge, E.-U., Yazdanpanah, M., and Heinrich, S. "Influences on the transition from bubbling to turbulent fluidization for Geldart's group B particles". In: *Powder Technology* 375 (2020), pp. 81—88.

Conference Contributions

Wytrwat, T., Hartge, E.-U., Yazdanpanah, M., and Heinrich, S. "Temperature Influence on Fluid Dynamics at the Transition from Bubbling to Turbulent Fluidization for Geldart's Group B Particles" *Fluidization XVI*. Guilin, China, 2019, Presentation.

Wytrwat, T., Hartge, E.-U., Yazdanpanah, M., and Heinrich, S. "Influence of Temperature on the Transition from Bubbling to Turbulent Fluidization for Geldart's Group B Particles". *Partec – International Congress on Particle Technology*. Nürnberg, Germany, 2019, Presentation.

Wytrwat, T., Hartge, E.-U., Yazdanpanah, M., and Heinrich, S. "Transition from Bubbling to Turbulent Fluidization for Geldart's Group B Particles". *23rd International Conference on FBC*. Seoul, Korea, 2018, pp. 642—651.

Wytrwat, T., Hartge, E.-U., Yazdanpanah, M., and Heinrich, S. "Fluid Dynamic Investigations on the Transition from Bubbling to Turbulent Fluidization using Geldart's Group B Particles". *ProcessNet Jahrestreffen – Fachgruppe Mehrphasenströmung*, Bremen, Germany, 2018, Presentation.

Wytrwat, T., Hartge, E.-U., Yazdanpanah, M., and Heinrich, S. "Capacitance probes for the investigation of the fluid dynamics of turbulent fluidized beds". *12th International Conference on Fluidized Bed Technology*, CFB 2017. Krakow, Poland, 2017, pp. 245—252.